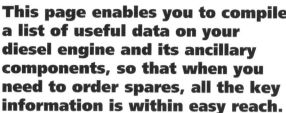
DATA BASE

This page enables you to compile a list of useful data on your diesel engine and its ancillary components, so that when you need to order spares, all the key information is within easy reach.

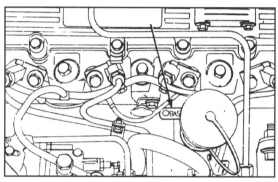

The engine number will be stamped on a plate fixed to the engine block.

The fuel pump should have a data plate fixed to its body.

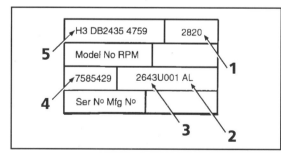

On most Lucas and Stanadyne pumps (for instance):
1 - max. no load engine speed
2 - fuel pump code
3 - engine maker's part no.
4 - pump serial no.
5 - model no.

Make of engine: ....

Model: ....

Power output: ....

Type of installation: ....

Engine number: ....

Operating key number (if appropriate): ....

Ancillary componants -

descriptions and part numbers: ....

....

....

....

....

....

....

....

....

....

Name, address and telephone number of local dealer:

....

....

....

....

....

First published in 1998 by: **Porter Publishing Ltd.**

The Storehouse
Little Hereford Street
Bromyard
Hereford HR7 4DE
England

Tel: 01 885 488 800
Fax: 01 885 483 012

British Library Cataloguing in Publication Data.

A catalogue record for this book is available from the British Library.

ISBN 1-899238-26-3

Series Editor: Lindsay Porter
Image scanning and origination: Mark Leonard
Design: Porter Publishing Ltd.
Layout and Typesetting: Pineapple Publishing, Worcester.
Printed in England by The Trinity Press, Worcester.

Perkins

A complete range of manuals is available, covering every model of
Perkins diesel engine. Please contact your local Perkins Power Centre
or, in case of difficulty, contact:

The Publicity Distribution Centre,
Perkins Engines,
Eastfield, Peterborough,
PE1 5NA, England

Tel: 01 733 567 474
Fax: 01 733 582 240

(Outside the UK, delete the first "0" of the above number, add "0044", then
dial the rest of the number given above.)

Porter Manuals

There is a full and growing range of Porter Manuals covering a large
number of cars and LCVs. Porter Publishing also publishes:

- manuals covering
  diesel engine-powered vehicles

- leisure manuals, on caravans,
  trailers, motor caravans, etc.

- over 60 classic car and motor sport
  video tapes

Please contact your local book shop or
auto-accessory store or, in case of difficulty, contact:

Porter Publishing,
The Storehouse, Little Hereford Street, Bromyard, Hereford, HR7 4DE
Tel: 01 885 488 800   Fax: 01 885 483 012

(Outside the UK, delete the first "0" of the above number, add "0044", then
dial the rest of the number given above.)

Diesel Engines Fault Finding

& Diagnostics Manual

by
David Ferguson

Acknowledgements

This book was produced at the request of Perkins Engines but is intended - again at their suggestion - to cover all diesel engine types. We are extremely grateful to Perkins Engines for all of the invaluable advice, assistance and practical information which they were able to supply, and for the generous support of a number of key individuals within that company. In addition, Hindle Power of Peterborough provided useful advice. The following companies have also kindly allowed use of illustrations: Chalywn Equipment, V L Churchill, Dieseltune, FIAT Auto UK, Ford UK, LDV, LucasVarity, Mercedes-Benz, Volkswagen Group.

The author would like to thank the following companies for their kind and generous assistance, advice and use of facilities: Mr A J Littlefield, Tri Service Resettlement Organisation, Aldershot, Hants, Mr J Gilbertson, British Waterways, Newark Workshop, Newark, Notts and Messrs K & S Butler, Newark Line River Cruisers, Newark, Notts.

CONTENTS

*Detailed Contents are shown
at the start of each chapter.*

ENVIRONMENT FIRST!

Keep it clean!

• *Maintaining a diesel engine in first-class condition is good for the environment, good for health - and good for the pocket!*

• *There's only one 'environment' on this world - so play **your** part in looking after it!*

Follow the Oil Care Code

• *When you drain your engine oil - don't oil the drain!* Pouring oil down the drain will cause pollution. It is also an offence.

• The used oil from the sump of just one car can cover an area of water the size of two football pitches, cutting off the oxygen supply and harming swans, ducks, fish and other river life.

• Don't mix used oil with other materials, such as paint or solvents, because this makes recycling very difficult.

• Take used oil to an oil recycling bank. Telephone FREE in the UK 0800 663366 to find the location of your nearest oil bank, or contact your local authority recycling officer.

CHAPTER 1
SAFETY FIRST!

It is vitally important that you always take time to ensure that safety is the first consideration in any job you do. A slight lack of concentration or a rush to finish the job can often result in an accident, as can failure to follow a few simple precautions. Whereas professional mechanics are trained in safe working practices others, without their training and experience, must find them out for themselves and act upon them. Remember accidents don't just happen, they are caused, and some of those causes are contained in the following list. Above all, ensure that whenever you work on an engine or its components you adopt a safety minded approach at all times. Always think ahead! Any danger may not be immediate but could happen later on with fatal consequences - especially so with marine engines.

Be sure to consult the suppliers of any materials and equipment that you may use, and to obtain and read carefully any operating and health and safety instructions or warnings that may be available on packaging or from manufacturers and suppliers. Always follow the safe working practices and observe all safety precautions given in the relevant Workshop Manual or Users' Handbook.

Important Points

DO NOT change the application or specification of an engine without seeking the manufacturer's advice. The output speed range and power of a vehicle engine may not match the requirements of a boat or generating set, nor may the mountings, lubrication or cooling system be suitable without some alterations. Changing the engine specification may be downright dangerous, especially if there is a risk of overspeed. Diesel engines can run on a wide range of fuels and may continue to run or even overspeed to the point of self destruction if oil can leak into the inlet manifold or past the piston rings. A nearby gas leak can have similar effects.

WHEN INSTALLING a generating set or boat engine, always follow the recommendations given by the manufacturer or contained in the manufacturer's Installation Manual and other relevant literature.

on, around, or underneath a raised vehicle unless axle stands are positioned under secure, load bearing underbody areas, or the vehicle is driven onto ramps. Engines and components should be securely mounted and supported on workbenches, stands or trestles. Unless dry docked, dried out or on a slipway boats can move unpredictably and without warning.

SEEK HELP if you need to lift something heavy which may be beyond your capability. There are right and wrong ways to lift items and a free guidance leaflet is obtainable from the Health and Safety Executive. If in doubt use lifting gear, hoists or jacks.

ALWAYS ensure that the safe working load of any lifting gear, hoists or jacks is sufficient for the job, and is only used as recommended by the manufacturer.

1. ALWAYS ensure that the equipment you are working on is properly supported when raised off the ground. Don't work

2. When lifting an engine, use the correct engine lift supplied by the manufacturer.

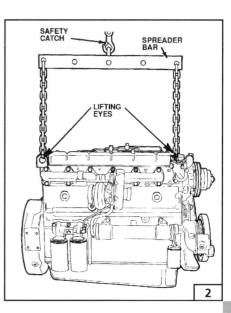

NEVER attempt to loosen or tighten any nuts, bolts or plugs that require a lot of force unless the vehicle, generating set or engine is properly supported and in safe condition. Wherever possible initially slacken tight fastenings before raising the vehicle off the ground or removing the engine from its mountings.

MAKE SURE that spanners and all other tools are the right size for the job and are not likely to slip. Do not use metric spanners on imperial fastenings or vice versa: close enough is not good enough. Never try to "double up" spanners to gain more leverage.

BE AWARE that certain generating sets, compressors etc. may start automatically or by remote control. Take care to ensure that any installed engine you may work on has all means of starting it *disabled* before you start work.

NEVER start the engine unless the load has been removed e.g. a vehicle should always have its gearbox in neutral. Ensure that a vehicle's handbrake is fully applied before the engine is started.

NEVER run catalytic converter equipped engines without the exhaust system heat shields in place.

TAKE CARE when parking vehicles fitted with catalytic converters. The "cat" reaches extremely high temperatures and any combustible materials under the vehicle, such as long dry grass or cleaning rags, could ignite.

TAKE CARE to avoid touching any engine or exhaust system component unless it is cool enough so as not to burn you.

BE METICULOUS and keep the work area tidy - you'll avoid frustration, work better and lose less.

ALWAYS keep antifreeze, coolant, brake and clutch fluid away from paintwork. Wash off any spills immediately.

CLEAN UP any spilt oil, grease, or water off the floor immediately, before there is an accident. Do not hose spillages into drains, use an absorbing compound.

NEVER drain oil, coolant or other fluids while they are hot. Allow them to cool sufficiently to avoid scalding you.

NEVER siphon fuel, coolant, cleaning agents, solvents or other liquids by mouth, or allow prolonged contact with the skin. There is an increasing awareness that they can damage your health. Best of all, use a suitable hand pump and wear protective gloves or apply a suitable barrier cream to your hands.

ALWAYS work in a well ventilated area and don't inhale dust - it may contain asbestos (e.g. from joints or gaskets) or other poisonous substances.

NEVER take risky short cuts or rush to finish a job. Plan ahead and allow plenty of time.

KEEP children and animals right away from the work area and from unattended vehicles, generating sets and engines. Remember that some equipment, especially generating sets, may start automatically and boat engines can often be started remotely.

ALWAYS wear eye protection when working under a vehicle or other machinery, also when using any power tools and where advised in the maintenance literature.

BEFORE undertaking dirty jobs or coming into contact with used engine oil or diesel fuel, use a barrier cream on your hands as a protection against skin damage and infection. Also, wear gloves intended for this purpose, available from most DIY outlets.

REMOVE your wrist watch, rings and other jewellery before doing any work on your engine, especially the electrical system. Never rest tools on the battery top - they could cause a short circuit and an explosion.

DON'T lean over or work on, a running engine unless strictly necessary, and keep long hair and loose clothing well out of the way of moving mechanical parts. Note that it is theoretically possible for certain types of lighting to make a rotating engine component appear to be stationary, so check! This is the sort of thing that happens when you're dog tired and not thinking straight. So don't work on your engine when you are overtired!

ALWAYS tell someone what you are doing and have them regularly check that all is well, especially when working alone.

ALWAYS seek specialist advice if you're in doubt about any job. The safety of your vehicle affects you, your passengers and other road users. The safety and reliability of your boat engine may save your life.

Fire

Only use flammable solvents in a well ventilated area, and disconnect the battery earth lead to prevent accidental electrical sparks which could ignite solvent vapours. Many solvent vapours are heavier than air and can accumulate in low areas such as bilges and inspection pits. If this occurs there is a real risk of explosion if the vapour is ignited.

3. Invest in a workshop sized fire extinguisher. Choose the carbon dioxide type or preferably dry powder, but never a water type extinguisher for workshop use. Water conducts electricity, and so may result in you getting an electric shock, it can also make worse a fuel oil or solvent based fire.

Be aware that fire can be a very, very serious hazard in a marine environment if only because the fire brigade may have trouble reaching you! So... take special care with marine engines and never 'cut corners' or take risks of any kind. Remember that fire will not only burn the fluids mentioned above, but also engine oil, nearby wooden and glassfibre structures, and may cause gas cylinders to explode!

Fumes

In addition to the fire dangers described previously, vapour from many solvents, thinners and adhesive is highly toxic and under certain conditions can lead to unconsciousness or even death if inhaled. The risks are increased if such fluids are used in a confined space so always ensure adequate ventilation when handling materials of this nature. Treat all such substances with care, always read the instructions and follow them implicitly.

Never have the engine running in any enclosed space unless the exhaust fumes are piped to the outside of the building or boat by leak-free piping. Always ensure that a vehicle is outside the workplace in open air if the engine is running. Diesel fumes contain poisonous carbon monoxide, even though it isn't present in such large quantities as in petrol engine exhaust fumes.

Inspection pits are another source of danger from the build up of fumes. Never drain fuel oil or use solvents, thinners adhesives or other toxic substances in an inspection pit as the extremely confined space allows the highly toxic fumes to concentrate. Running the engine over the pit can have the same effect, as can a nearby gas leak. Note that even gases like nitrogen and carbon dioxide can be dangerous, since although they are not poisonous, they are asphyxiants and, in sufficient quantities, can suffocate you.

Cooling

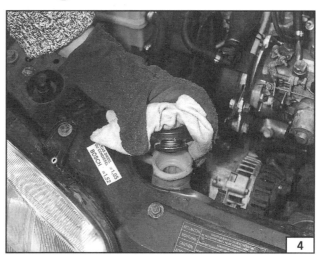

4. In order to raise the boiling point of the coolant, the cooling system is kept under pressure when the engine is hot or even warm. This means that the depressurisation of a hot or warm system is likely to result in boiling over. The system is depressurised if any part of it is opened, usually the radiator, heat exchanger, or catcher tank pressure cap. Therefore, there is a very real danger of scalding if the system is opened when it is hot and under pressure, as boiling coolant will erupt or spray from the opening. If in doubt, wait until the engine is cool and use appropriate protection.

Mains Electricity

5. Best of all, use rechargeable tools and a 12v DC inspection lamp powered from a remote battery - both are much safer! However, if you do use mains powered equipment, ensure that the appliance is wired correctly to its plug, that where necessary it is properly earthed (grounded), and that the fuse is of the correct rating for the appliance concerned. Do not use any mains powered equipment in damp conditions or in the vicinity of fuel, fuel vapour or the vehicle battery.

Before using any mains powered electrical equipment take one more simple precaution, use an RCD (Residual Current Device) circuit breaker. Then if there is a short, the RCD circuit breaker minimises the risk of electrocution by instantly cutting the power supply. Simply fit the RCD into your electrical socket before plugging in your electrical equipment.

The Battery

Never cause a spark, smoke or allow a naked flame near a battery, even in a well ventilated area. A certain amount of highly explosive hydrogen is given off as part of the normal charging (and jump starting) process. You can help to avoid sparking by switching off the battery charger's power supply before the charger leads are connected or disconnected. Battery leads should be shielded, since a spark can be caused by any electrical conductor which touches its terminals or exposed connecting straps.

Before working on the fuel or electrical systems, always disconnect the battery earth (ground) terminal.

When charging the battery from an external source, disconnect both battery leads before connecting the charger. If the battery is not of the "sealed for life" type, loosen the filler plugs or remove the cover before charging. For best results the battery should be given a low rate "trickle" charge. Do not charge at an excessive rate or the battery may burst.

Always wear gloves and goggles when carrying or topping up a battery. Even in its diluted form (as it is in the battery) the electrolyte acid is extremely corrosive and must not come into contact with eyes, skin or clothes.

If battery acid is splashed into the eyes or onto skin or clothing, wash off IMMEDIATELY with copious amounts of clean water. Seek medical attention if necessary.

If you wish to jump start an engine from the battery of another vehicle or a separate battery, be sure to connect the jump leads in the correct order to avoid potential damage to the batteries and electrical components.

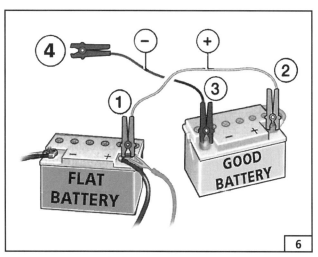

6. • Connect one end of the positive (+ and usually red) cable to (+) the terminal of the flat battery (**1**).
• Connect the other end of the (+) cable to the (+) terminal of the good battery (**2**) - which must be of about the same capacity as the flat battery.
• Connect one end of the negative (- and usually black) cable to the (-) of the good battery (**3**).
• The other end of the (-) cable can be connected to the negative terminal on the flat battery, or a metal part of the engine block (**4**).

But be careful with this method! Modern vehicles and engines use a negative earth system but some older ones may use a positive earth. Unless you are sure, it is safer to connect only between the battery terminals.

IMPORTANT NOTES:
• **Do not cause any shorting, especially with jump lead cables or clips.**
• **Make sure the jump lead cables can't touch any moving or hot parts on the engine.**
• **Keep away from both batteries, to avoid risk of acid burns.**
• **Keep sources of ignition, such as cigarettes or flames, away from both batteries.**

Run the engine of the donor vehicle (if appropriate) at a fast idle while jump starting, to avoid rapidly flattening the donor battery. When the engine has been successfully started, disconnect the jump leads in the reverse order to that of connection, and do so only when both engines have been returned to a normal tick-over, or the donor engine has been switched off.

Engine Oils

Take care and observe the following safety precautions when working with used engine oil. Apart from the obvious risk of scalding when draining hot oil from an engine, there is a risk to health from the contaminates that are contained in all used oil.

Always wear disposable plastic or rubber gloves, intended for this purpose, when draining oil from your engine.

There are very real health hazards associated with all used engine oil - and used diesel engine oil contains particularly dangerous chemical compounds. According to one expert source, "Prolonged and repeated contact may cause serious skin disorders including dermatitis and cancer". Use a barrier cream on your hands and try not to get oil on them. Where practicable, wear protective gloves and wash your hands with hand cleaner soon after carrying out the work. Keep oil out of the reach of children.

NEVER EVER dispose of old engine oil into the ground or down a drain. In the UK and in most EU countries, every local authority must provide a safe means of oil disposal. In the UK try your local Environmental Health Department for advice on waste disposal facilities.

OIL POLLUTES WATER USE YOUR BRAIN - NOT THE DRAIN!

Fuel Injectors

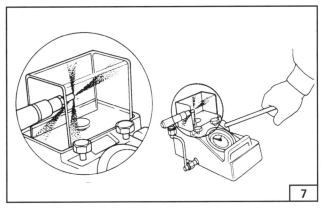

7. The injectors are at the heart of engine performance, and it's likely that at some stage you are going to want to test them. It must be stressed that specialist equipment is needed to test an injector properly, and although some claim that it is possible to observe the spray pattern by removing the injector from the engine and running it connected to its supply pipe from the fuel pump, with its nozzle contained within a glass jar, this is not a very useful test and is potentially very dangerous. Fuel is injected at very high pressures, enough to pierce skin and enter the bloodstream, which could be fatal. Diesel fuel is also harmful to the skin. You should protect your hands with barrier cream and wear protective gloves where practicable before opening any part of the fuel system. Unit injectors should always be tested, repaired and reset with special equipment designed and supplied for the purpose.

Plastic Materials

Work with plastic materials brings additional hazards into workshops. Many of the materials used (polymers, resins, adhesives and materials acting as catalysts and accelerators) readily produce a very dangerous situation in the form of poisonous fumes, skin irritants, risk of fire and explosions. Do not allow resin or two pack adhesive hardener, or that supplied with filler or two pack stopper to come into contact with skin or eyes. Do not allow anaerobic locking compounds (e.g. "Loctite" products) and silicone sealing compounds to be used near to each other. They do not even have to touch for a chemical reaction producing toxic fumes to take place.

Lifting equipment, hoists, jacks, (axle) stands and ramps

Throughout this book you will find references to lifting equipment, jacks, axle stands and similar equipment, and we make no apologies for being repetitive! This is one area where safety cannot be overstressed - your life could be at stake!

Special care must be taken when using any type of lifting equipment: Overhead cranes, hoists and jacks are intended for lifting purposes only, not for supporting an engine, vehicle or anything else whilst it is being worked on. Full details of safe working practices should be found in the Workshop Manual for your engine.

Asbestos

8. Asbestos dust causes lung disease and lung cancer. Although the use of asbestos has been reduced, and in some engines has been eliminated, it may be encountered in old engines or replacement parts. The risk from new components is low and any containing asbestos should be clearly labelled. The greatest danger is from dust produced during the removal of old gaskets and joints. Sensible precautions are to:

• Work in an area with good ventilation.

• Wear an efficient particle mask.

• Use a hand scraper to remove the old material; do not use a rotary wire brush

• Ensure that the material to be removed is wet with oil or water to contain loose particles.

• Wipe off all gasket dust and particles from the work area, never blow it off with compressed air.

• Spray all asbestos dust with water and dispose of old gaskets, dust and particles in a sealed plastic bag.

• Dispose of asbestos waste in accordance with your local authority regulations.

• Wash your hands thoroughly after you have finished working and certainly before you eat, drink or smoke.

IMPORTANT NOTE: The above information ONLY relates to engine-related components. Brake components, for instance, should NEVER be sprayed with oil or water.

Fluoroelastomers

MOST IMPORTANT! PLEASE READ THIS SECTION!

If you maintain and service your engine in the normal way, none of the following may be relevant to you, unless you encounter an engine or engine component which has been on fire (even in a localised area), or subject to heat, in say a breaker's yard or if any second hand parts have been heated in any of these ways.

Many synthetic rubber-like materials used in engines contain a substance called fluorine. These materials are known as fluoroelastomers and are commonly used for oil seals, wiring and cabling, bearing surfaces, gaskets, diaphragms, hoses and 'O' rings. Seals containing fluoroelastomers are also known at Viton seals. If they are subjected to temperatures greater than 315 degrees C, they will decompose and can be potentially hazardous. Fluoroelastomer materials will show physical signs of decomposition under such conditions in the form of charred black sticky masses. Some decomposition may occur at temperatures above 200 degrees C, and it is obvious that when an engine has been in a fire or has been dismantled with the aid of a cutting torch or blow torch, the fluoroelastomers can decompose in the manner indicated above.

In the presence of water or humidity, including atmospheric moisture, the by-products caused by the fluoroelastomers being heated can be extremely dangerous. According to the Health and Safety Executive, "Skin contact with this liquid or decomposition residues can cause painful and penetrating burns. Permanent irreversible skin and tissue damage can occur". Damage can also be caused to the eyes or by the inhalation of fumes created as fluoroelastomers are burned or heated.

After fires or exposure to high temperatures observe the following precautions:

• Do not touch blackened or charred seals or equipment.

• Allow all burnt or decomposed fluoroelastomer materials to cool down before inspection, investigations, tear-down or removal.

• Preferably, don't handle parts containing decomposed fluoroelastomers, but if you must do so, wear goggles and PVC (polyvinyl chloride) or neoprene protective gloves. Never handle such parts unless they are completely cool.

• Contaminated parts, materials, residues and clothing, including protective clothing and gloves, should be disposed of by an approved contractor to landfill or by incineration according to national or local regulations. Oil seals, gaskets and 'O' rings, along with contaminated material must not be burned locally.

• If there is contamination of the skin or eyes, wash the affected area with a continuous supply of clean water or with calcium hydroxide solution for 15-60 minutes. Obtain immediate medical attention.

Workshop Safety - summary

• Always have a fire extinguisher of the correct type at arms length when working on an engine - this is especially important when working on the fuel system.

If you do have a fire, DON'T PANIC. Use the extinguisher effectively by directing it at the base of the fire.

• NEVER use a naked flame near petrol, paraffin (kerosene), diesel fuel, thinners or anywhere in the workplace.

• NEVER use petrol (gasoline) to clean parts. Use paraffin, white (mineral) spirits or a proprietary bio-degradable cleaning agent instead. Be aware that paraffin and white spirits can cause some rubber components to perish. Do not dispose of solvents or cleaning materials down drains unless the supplier's instructions indicate otherwise.

• NO SMOKING! There's a risk of fire or transferring dangerous substances to your mouth and, in any case, ash falling into mechanical components is to be avoided!

• BE METHODICAL in everything that you do, use common sense, and think of safety at all times.

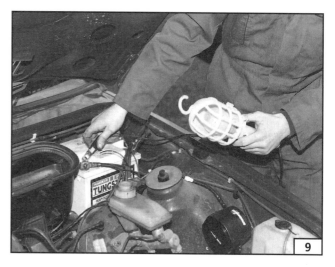

9. KEEP your inspection lamp away from any source of flammable or combustible substances.

Please be sure to read the whole of CHAPTER 1, SAFETY FIRST! before carrying out any work on your car.

CHAPTER 2

HOW IT WORKS

CHAPTER 2
HOW IT WORKS

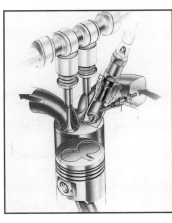

Most people can recognise a diesel engine from its clatter on tick-over and the... shall we say, distinctive exhaust pipe output! But ask them what makes it tick (or clatter, as the case may be) and it's not so easy - even though the first stage in fixing a fault is understanding it!

In this chapter, we aim to explain what it is that makes a diesel engine work, as well as some of the advantages of Professor Diesel's brainchild, over the more familiar petrol engine.

What is a Diesel Engine?

1. Diesel engines are very similar in construction to petrol engines and also operate on either the two- or four-stroke cycles. However, while two-stroke petrol engines are encountered on small lightweight applications, such a mopeds and portable equipment, two-stroke diesels are almost exclusively used for very large slow-speed applications such as ship engines. Since readers of this book are unlikely to encounter two-stroke diesels we will not be discussing them further.

The principle difference between diesel and petrol engines is that a diesel engine does not need an ignition system (spark plugs, coil etc.) to fire it. Instead, intake air within each cylinder is compressed by the piston to the point where it becomes hot enough to ignite the fuel.

2. You may be surprised that the simple act of compressing air can make it so hot, but it's certainly no secret to cyclists that the act of compressing air heats it up, which is why many bicycle pumps can become too hot to handle!

But back to the diesel engine: When the piston is at the top of its compression stroke, and the air is hot (about 700 degrees C, or more, depending on the engine) diesel fuel is injected into the cylinder, whereupon it is ignited by the heat in the air.

The diesel fuel-and-air mixture ignites and expands with almost explosive power, forcing the piston back down the cylinder with sufficient energy to do useful work and enough "shove" to compress the next charge of air right up to 700 degrees C again. It's a wonderfully simple system, and it does away with any possible unreliability in an electrical ignition system.

But apart from the elimination of spark plugs, the diesel engine has other advantages. The greatest of these is that by compressing its intake air to a much greater extent than a petrol engine (a typical compression ratio of around 14:1 on large engine, to about 22:1 on small, modern engines), the diesel is more "thermally efficient". This means that it more efficiently produces power from a given quantity of fuel. The result: a vehicle that travels many more miles or an engine that runs many more hours per gallon than its petrol fuelled equivalent.

Diesel History

It was in 1890 that Doctor Rudolph Diesel developed the theories of the "economical thermal motor" which, through its high cylinder compression, significantly improved efficiency. Although he was the first to file a patent for this "compression-ignition" engine, an engineer called Ackroyd Stuart had previously had similar ideas. He proposed an engine in which air was drawn into the cylinder, compressed, and then forced - at the end of the compression stroke - into a bulb into which fuel was sprayed. To start the engine, the bulb was heated externally by a blow lamp, and once started kept itself running without external heat.

Ackroyd Stuart wasn't concerned with the advantages of operating very high compression pressures, he was simply experimenting with the possibility of avoiding the use of spark plugs, so he missed out on the greatest advantage - fuel efficiency. This is why the term "diesel engine" has remained, as it is Dr. Diesel's theories which form the basis of the modern "compression-ignition" engine.

The Cycle of Operations

This is the full four-stroke cycle of diesel engine operation.

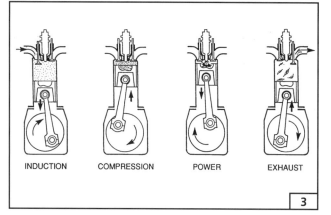

INDUCTION   COMPRESSION   POWER   EXHAUST

3

3. On the first (downward, induction) stroke of the piston a fresh charge of air alone is drawn into the cylinder through the open inlet valve.

On the second (upward, compression) stroke, the inlet and exhaust valves are closed and the air is compressed into a volume typically 17 times smaller than the total cylinder volume, so becoming very hot.

Just before the start of the third (downward, power) stroke, fuel is injected into the combustion chamber through a nozzle. During injection the fuel is "atomised" into tiny particles which mix evenly with the compressed air to make a combustible self-igniting charge. Energy is unleashed by combustion as the piston starts its decent on the power stroke. Injection continues, causing the burning fuel to maintain constant pressure on the piston.

The exhaust valve opens as the piston starts its fourth (upward, exhaust) stroke, exhaust gases exiting through the valve.

Diesel Engines Ancient and Modern

4. This 45 bhp four cylinder Wolf Engine is the oldest Perkins engine still in working order. It was despatched 2nd May 1934, only two years after the company was originally founded in 1932.

4

After a worldwide search, the 'Wolf' was discovered, still being operated by a Southampton Company to power a standby generator set.

Originally installed in a Carrimore truck, it was returned to Perkins in 1939 for a major overhaul. The engine dropped from sight during the war period but re-emerged when it was discovered installed in a Daimler taxi in 1954.

In the late 1960's it was set aside where it did nothing until 1973, when it was overhauled by Mr Musson's own workshop and installed in a standby generating set. The engine was found still in active service, swapped for a modern diesel and brought back to its birthplace at Peterborough.

5

5. Today's light diesels have been so extensively refined however, when fitted to a car and once on the move, it is often very difficult to tell whether you're driving a diesel or a petrol... sometimes only the greater flexibility of the diesel engine (or an open window) gives the game away. This is a 1997 VW Passat 1.9 litre TDI engine, delivering 110 bhp.

Similarities with their forebears are even more apparent with industrial diesel engines, which are less shrouded in cosmetics.

Advantages and Disadvantages of the Diesel Engine

The petrol engine is comparatively inefficient, capable of converting only about 26% of the energy in its fuel into useful work. The diesel engine, however typically returns a 36% fuel efficiency, and so is more economical to run. For UK engines not used in road vehicles, there is yet one more advantage: there is less tax on the fuel!

The elimination of an electrical ignition system is an obvious advantage for all types of use, but in a boat or on a building site, the increased reliability, in what are, after all, very damp and unfriendly environments, is even more important. The diesel engine also provides high torque (pulling power) over most of its speed range, which makes a diesel engined vehicle much more flexible to drive than its petrol engined counterpart. It is an advantage too in marine engines since the high torque at lower speeds makes it easier to use the engine's power effectively.

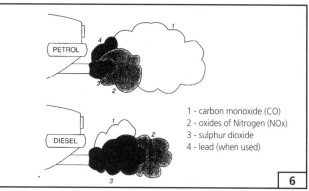

PETROL

DIESEL

1 - carbon monoxide (CO)
2 - oxides of Nitrogen (NOx)
3 - sulphur dioxide
4 - lead (when used)

6

6. And the advantages continue. The diesel's exhaust gases are relatively harmless compared with those of the petrol engine. Carbon monoxide (CO) is virtually non-existent in the diesel's emissions, so the only undesirable gases present in any significant quantities are hydrocarbons (HC) (not shown in this chart), oxides of nitrogen (NOx), and soot (or particulates) in the form of black smoke. These have been accused of causing asthma and lung cancer - and the greatest offenders here are heavy commercial vehicle diesels, such as trucks and buses, which are often old and poorly maintained.

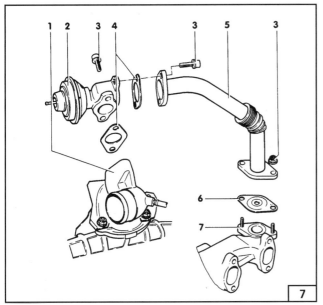

7

7. NOx can be reduced by Exhaust Gas Recirculation. This system takes some of the exhaust gas from the exhaust manifold (**7**) via duct (**5**) to the intake manifold (**1**). The process is controlled by valve (**2**), and by lowering combustion temperature, the production of NOx is reduced. (Illustration, courtesy V.A.G.).

8. Oxidation-type catalytic converters are used to substantially reduce hydrocarbons and the remaining CO. As for the remaining soot, improvements in fuel injection and combustion combined with exhaust system particulate extractors are rapidly eliminating it. Good maintenance of diesel engines helps to keep black smoke down to a minimum.

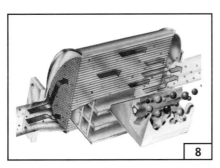

8

This is a cutaway of the diesel's oxidation catalytic converter, which reduces the emissions of CO, hydrocarbons and the attendant nasty diesel smell. (Illustration courtesy V.A.G.).

Another important aspect - and it's safety-related - is that diesel fuel isn't volatile (it doesn't vaporise easily) so the likelihood of fires is much smaller with diesels - particularly as there is no ignition system to start them! For marine installations, and those in other enclosed areas, this is a very significant advantage as the dangers of an explosive petrol/air mixture forming in the bilges of a boat or the low areas of buildings are very real.

Of course there are disadvantages, among them the clattering noise that most diesels suffer from when ticking over, and the greasiness of the fuel. But these are, mainly, only significant to owners of diesel-engined cars and are really small penalties to set against the substantial gains.

Diesel Engine Construction

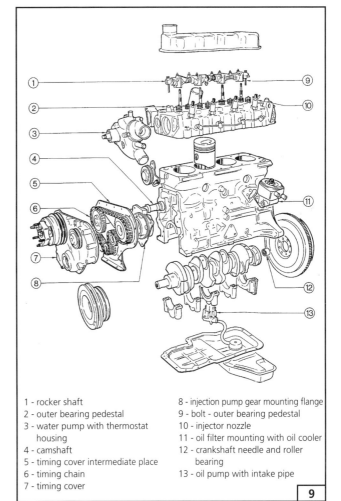

1 - rocker shaft	8 - injection pump gear mounting flange
2 - outer bearing pedestal	9 - bolt - outer bearing pedestal
3 - water pump with thermostat housing	10 - injector nozzle
4 - camshaft	11 - oil filter mounting with oil cooler
5 - timing cover intermediate place	12 - crankshaft needle and roller bearing
6 - timing chain	13 - oil pump with intake pipe
7 - timing cover	

9

9. The diesel engine's basic construction is similar to that of the petrol engine as this exploded view shows. However, similar components are usually heavier and more robust to withstand the much higher compression pressures used in the diesel engine.

10A. Piston crowns, however, are specially designed to suit diesel combustion, and often (although not always), in order to achieve the highest compression ratios, the piston tops are higher than the top face of the cylinder block when each piston is at the top of its stroke. (Illustration courtesy Ford Europe Ltd).

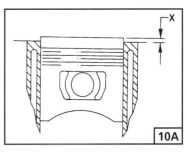

10A

10B. In many cases, pistons crowns contain a combustion chamber within them, such as in this Perkins 3000 series piston.

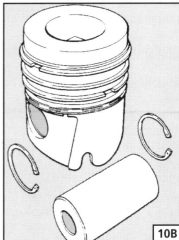

10B

10C. The compression ratio is the difference between the volume 'X' above the piston when it is at the bottom of its stroke (BDC) compared with the volume 'Y' when it is at the top of its stroke (TDC). (Illustration courtesy Ford Europe Ltd.).

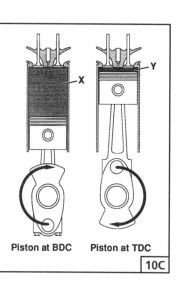

Piston at BDC   Piston at TDC

10C

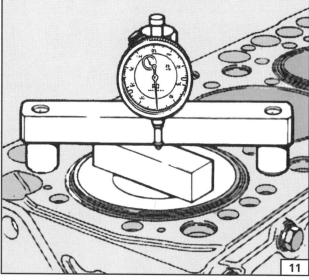

11

11. The pistons used on smaller diesel engines are almost invariably designed to protrude above the top of the cylinder block face when at top dead centre (TDC). As an engine is built up, the amount of protrusion must be checked and correctly set, if outside the manufacturer's recommended tolerances. The amount of piston projection is critical to provide the correct compression ratio, while assuring that the valves do not come into contact with the piston crowns. The height is determined by rotating the engine, by hand, slowly past the TDC position, and noting the maximum height measurement on the dial gauge. (Illustration, courtesy Ford Europe.)

12. With some engines, such as the Perkins 100 Series, and many other small engines, a range of gaskets in different thicknesses is available. In some cases, there are notches cut out of their edges to distinguish between the different thicknesses. An appropriate thickness is selected to give the correct effective protrusion - above the top

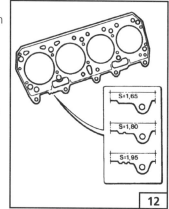

S=1,65

S=1,80

S=1,95

12

face of the gasket, when fitted, rather than the top of the block, of course. Check with your engine's manufacturer's manual for the gasket thickness identification system used.

For other engines, a selection of pistons may be available. Then, it is a mater of fitting a set of pistons, measuring their heights and selecting different sizes of piston, as necessary, in order to obtain the correct piston height. In other cases, especially on older, larger engines, the piston crowns may have to be machined if the protrusion is too high (although what should be done if the pistons are too low is not always made clear by the manufacturer!).

13

13. The valve gear is also conventional, as is the camshaft drive arrangement, with the exception that it also drives a fuel injection pump on some engines. The drive is normally by toothed rubber belts, chains or gears.

The fuel injection pump is driven by the idler gear, which usually also drives the camshaft.

The major differences between diesels and petrol engines lie in the air intake system, which has no throttle butterfly (see *Injection Pump Governor*, later in this chapter), the design of the combustion chambers and the presence of an injection pump or unit injectors in place of an ignition distributor and carburettor or fuel injection system. In the 'traditional' fuel injected petrol engine, petrol is injected into the manifold at low pressure and mixes with the air before entering the cylinders. In a diesel (and in some state-of-the-art petrol engines), fuel is injected at very high pressure directly into the cylinders.

14. Most diesel engines are of the direct injection (open chamber) type. These have a relatively simple flat cylinder head with the combustion chamber formed within the piston crown - note the important swirl imparted to the

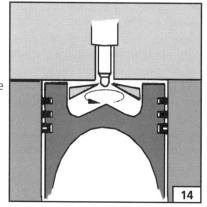

14

incoming air by the design of the inlet port. These engines are reliable cold starters and return good fuel consumption but are inclined to noise and roughness in operation, and to incomplete combustion, producing black smoke from the exhaust. Direct Injection engines invariably use multi-hole nozzle injectors in order to assist fuel distribution throughout the combustion chamber.

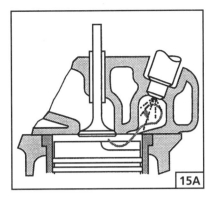

15A. Because they have had to compete with petrol engines, most diesel car engines have traditionally been of the indirect injection type, where combustion starts in a pre-combustion ante-chamber. Again note the swirling of the compressed intake air in the pre-combustion chamber. The pre-combustion chamber, one for each cylinder, is contained within the cylinder head, with an injector protruding into it. These engines do not give such good fuel consumption as the directly injected type, nor are they such reliable cold starters. However, they are quieter and smoother in operation which is an essential requirement of a diesel car engine.

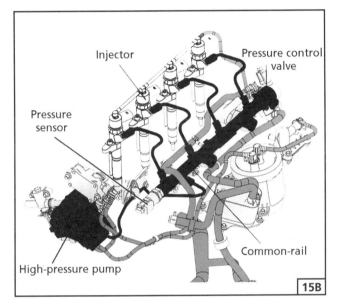

15B. In an attempt to provide the best of both worlds, latest developments are of the common rail type of injection system which differs in several ways from other direct fuel injection systems. Whereas conventional systems regenerate the pressure for each injection anew each time, in the case of the new system, the fuel pressure is maintained in a common rail and distributed to the injection nozzles. The electronic engine control regulates the high pressure of up to 1,350 bar independently of the injection sequence according to the speed and load of the engine. The injector nozzles, which are equipped with special solenoid valves, can also be controlled in a variable manner as required.

Together, the high injection pressure, which is also available at low speeds, and the variable injection process control produce much improved carburation in the cylinders. The result is greatly improved fuel consumption and lower exhaust emissions.

If you want to know more about the reasons why the two main types - direct and indirect injection - exist, and their exact differences, you'll find an explanation under the *Injectors* section later in this chapter.

Diesel Engine Lubricating Oil

There are many engine oils on the market, some being specific to diesel engines while others are suitable for both petrol and diesel engines.

When choosing which oil to use, always begin by reading the recommendations in the literature applicable to your engine. This is most important, since a lubricant's requirements will vary with the application of the engine. The literature will give either an API (American Petroleum Institute) or ACEA specification - more of which shortly. But first we'll look at the ingredients of modern engine oils.

Mineral oils: are all similar until special additive packages are 'stirred in'. The 'package' comprises a basic group of important additives; the proportions and exact nature of each depending upon the engine's application. This group includes:-

a) **'Detergents':** usually metallic compounds that keep engine components clean and prevent deposits from forming, particularly on pistons and rings. They control the problem of 'varnish' formation, which is a coating created from oil oxidation products that bake onto engine components during high temperature use.

b) **'Dispersants':** non-metallic compounds often used in conjunction with detergents. Dispersants are vital for keeping soot and other undesirable deposits from settling out on to engine surfaces, they keep the contaminants harmlessly in suspension. 'Sludge' - a black treacle-like emulsion of water, combustion products and oil formed under cold running conditions, is also combated by dispersants.

c) **'Anti-oxidants':** prevent the oil from degrading with increasing temperature, and so lengthen its life.

d) **'Anti-wear agents':** particularly important to minimise wear on rubbing surfaces such as pistons/cylinders, cams/followers and plain bearings.

e) **'Corrosion inhibitors':** protect metal surfaces against chemical attack by water which causes rust, and acids which may form as oxidation products in a deteriorating oil, or may be introduced into the oil as by-products of combustion. Corrosion leads to wear.

f) **'Anti-foam additives':** are needed because air in the engine crankcase is whipped into the oil by movement of oil and components. Aerated oil cannot perform properly as a lubricant and so aeration must be suppressed.

g) **'Pour point depressant':** engine oil must remain reasonably thin when cold so that it can circulate properly and not impose too much of a load on the starting system; this additive prevents oil thickening at low temperatures.

h) **Viscosity index (VI) improver:** this reduces the tendency of an oil to change its viscosity (thickness) as temperature varies. A multigrade oil is one containing VI improvers that make it suitable for use over a much wider range of temperatures than a single grade oil. Vehicle engine oils experience wide temperature variations, since the difference between a cold start on a winter morning and a long motorway thrash on a hot day is vast. However, marine or industrial engines suffer much less extreme variations so always check the manufacturer's recommendations. The international SAE multigrade classifications for oils comprise a cold viscosity rating, indicated by a 'W' for 'winter' (e.g. '10W'), and a hot viscosity rating (e.g. 40). The higher the number, the thicker or more viscous the oil. Engine oil viscosity requirements have generally been reduced over the years from 20W 50, through 15W 40 and currently to 10W 30 on some modern engines.

Which Multigrade Rating is Best?

The advantages of a thinner oil are many, and include better fuel economy due to decreased friction, easier, more reliable cold starting for much the same reason, and lighter smaller and cheaper starter motors and batteries. While a manufacturer will specify a particular grade, this will be ideal for the engine's new, perfect, state, but not necessarily ideal later in its life. In some cases where an engine is displaying certain signs of 'old age', such as bearing noise on start up, or low compression, it may prove advantageous to try a higher viscosity oil, since the greater viscosity can help restore cylinder compressions and affords greater protection to sloppy bearings.

Modern oils should be used with caution in old engines that were first used before more recent advances in oil technology and may not have been used for a long time. Sludge deposits in very old engines can be loosened by modern detergent oils, the loosened sludge blocking narrow oilways with disastrous consequences.

Why Diesel Engine Oil is Different

While petrol has a sulphur content around 0.03%, diesel fuel, at about 0.2%, has a considerably higher content. Sulphur is undesirable in a fuel because its combustion produces corrosive sulphuric acid and environmentally nasty sulphur oxides in the exhaust. Because of these higher levels, diesel lubricating oil has more harmful acids dumped in it and therefore requires more detergents and a surplus of alkalinity to help neutralise the acids. Incidentally, this production of acids is one of the major factors forcing the frequent oil change requirements of the diesel. Another is dilution of the oil by fuel , which in the case of petrol engine evaporates from the oil, something which diesel fuel cannot readily do.

The diesel's production of soot is the other major factor determining the difference between lubricating requirements. Diesels tend to burn 'sooty', so greater quantities of carbon deposits find their way into the engine oil. This unwanted internal effect must be combated by a different and stronger cocktail of dispersive agents within the oil, compared with that of petrol engine oil. The excess carbon held in suspension in the diesel's oil builds up in time to thicken it, in direct opposition to the effects of unburned fuel which thins it. And the oil's detergents are not ever-lasting, but get used up just like we consume our own detergents in the bathroom, so eventually old oil can be depleted of these vital agents. As you can see, therefore, the oil's design properties and additive balances can soon be lost - hence the importance of changing engine oil at the correct intervals. Other additive requirements for both diesel and petrol engine oils remain much the same.

What the Codes Mean

The oil's performance capabilities are specified by an API or ACEA coding somewhere on the package. The API codings for diesel engines are: 'CC', through to 'GG' (and beyond, as time goes by). The higher the second letter, the better the performance of the oil. As a guide, the more modern the engine, the further 'along the alphabet' the 'C' rating should be, and turbocharged engines need the highest spec. of all.

ACEA specification codes, which cannot be directly translated into API gradings, are rated 'D', followed by a number; or 'PD' followed by a number: the higher the number, the higher the spec.. It may not be too relevant to quote the exact performance specification for each code here; suffice to say that your User's Handbook will provide the oil performance code required, as well as the viscosity range, and it's simply up to you to find an oil that matches the manufacturer's recommendation.

Minerals vs Synthetics: Another choice can be made, however, and that is between mineral and synthetic oils, or the semi-synthetic type that slips in between the two. Many people assume that synthetic oil represents the pinnacle of engine protection and that mineral oils are poor relations. To an extent this is true, since synthetic oils - the products of chemical companies rather than oil companies, at dedicated chemical plants rather than refineries - offer obvious advantages. These include a longer life due to better oxidation stability (leading to a cleaner engine), a higher viscosity index (i.e. more stable viscosity over varying temperatures), better thermal conductivity (improved heat dispersal results in more efficient oil cooling), and better residual lubricity due to the longer adherence of oil to engine surfaces, providing more protection when the engine is started. (Owners and operators of little-used marine engines and standby generating sets please note!) Its decreased volatility (relative to that of its mineral cousin) also implies that oil consumption may be reduced.

So, generally speaking synthetic oils are better, but there are various qualities of synthetic oil available, just as there are on the mineral front, and some of the better mineral oils - subjected to high levels of refining such as 'hydro-cracking' - exhibit very high resistance to high temperatures and oxidation. The cost of an oil is generally a good indication of its quality.

Saving on Synthetics

For many engine applications though, the benefits of a good synthetic oil may not justify the high price, for those benefits are at a prime under rigorous operating conditions and in circumstances where maintenance may not be carried out at ideal intervals. One great advantage of synthetic oil is its resistance to 'coking' (more easily understood if it is referred to as 'cooking'!) which is a boon in a turbocharged engine, where the turbocharger oilways reach such high temperatures that they 'cook' mineral oil to form varnish and carbon deposits. This appreciably reduces the life of turbocharger bearings, especially where oilways are obstructed by carbon build-up.

A handy half-way measure is presented by the semi-synthetic oil, a blend of mineral and synthetic representing a mid-point in performance. Its reason for being is that it is, little more expensive than a good mineral oil, yet offers some of the advantages of synthetics - reduced volatility and improved high temperature performance being among them. Semi-synthetics do have a real role to play, and are often the best compromise between cost and performance.

The Fuel Injection Pump

The functions of the petrol engine's carburettor and ignition distributor can loosely be regarded as being replaced by the fuel pump. Why? Because it accurately measures the amount of fuel going into the engine (as does a carburettor), it determines the fuel injection timing (just as the ignition distributor sets the spark timing), and it distributes fuel to each of the engine cylinders (as the distributor cap and HT leads distribute electric current to the spark plugs!).

16. The fuel injection pump is a mechanical engine-driven pump, which takes its drive from the same chain, gears or belt that drives the engine camshaft. This is the Perkins-built Rover Maestro/ Montego diesel engine's belt drive.

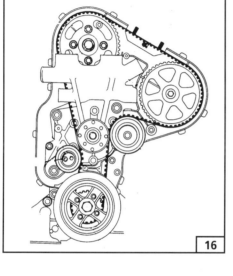

The types of fuel pump in common use are the distributor pump, the in-line pump and the unit injector system.

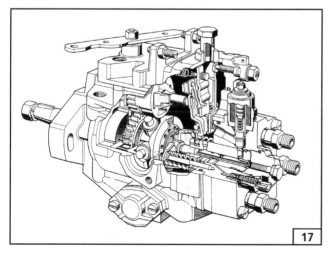

17. A distributor pump has one high-pressure element to serve all the engine's cylinders.

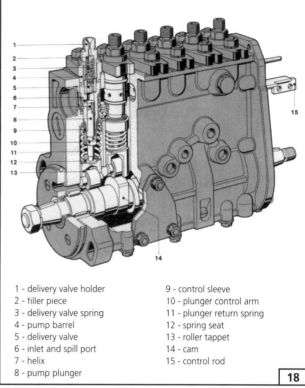

1 - delivery valve holder	9 - control sleeve
2 - filler piece	10 - plunger control arm
3 - delivery valve spring	11 - plunger return spring
4 - pump barrel	12 - spring seat
5 - delivery valve	13 - roller tappet
6 - inlet and spill port	14 - cam
7 - helix	15 - control rod
8 - pump plunger	

18. The in-line pump and unit injectors have separate high-pressure pumping elements dedicated to each cylinder. (Illustration courtesy of V.A.G.)

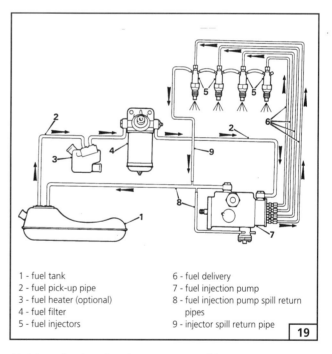

1 - fuel tank	6 - fuel delivery
2 - fuel pick-up pipe	7 - fuel injection pump
3 - fuel heater (optional)	8 - fuel injection pump spill return
4 - fuel filter	pipes
5 - fuel injectors	9 - injector spill return pipe

19. Most diesel engines have a separate lift pump which supplies fuel from the fuel tank to the injection pump. On some small diesel engines no lift pump is used, the fuel injection pump drawing fuel all the way from the tank. Because this is a suction system, it is important not to allow any air to be drawn in via any loose connections in the fuel lines, as air will impede injection, and in sufficient quantities can bring the engine to a complete halt.

The Distributor-type Injection Pump

This particular distributor pump's single element is operated by a cam ring or cam plate which has one cam per engine cylinder. It also, of course, has one injection outlet per cylinder.

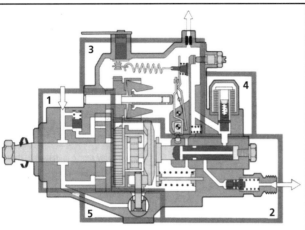

1 - vane-type supply pump. Supplies fuel from tank to injection pump cavity.
2 - high-pressure pump with distributor. Produces injection pressure, moves and distributes fuel to cylinders.
3 - mechanical governor. Controls engine speed, varies fuel delivery over control range.
4 - electromagnetic shut-off valve. Interrupts fuel delivery to stop engine.
5 - injection timing unit. Adjusts beginning of injection according to engine speed.

20

20. All distributor-type injection pumps feature the following internal functions: suction-type fuel-supply (transfer) pump, injection pressure pumping with fuel distribution to injectors, speed governing, injection timing variation, and engine shut-off. (However, distributor pumps fitted to generator sets, designed to run at a constant speed, usually have fixed injection timing, without an injection timing advance facility.)

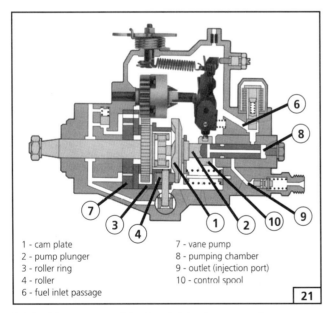

1 - cam plate
2 - pump plunger
3 - roller ring
4 - roller
6 - fuel inlet passage
7 - vane pump
8 - pumping chamber
9 - outlet (injection port)
10 - control spool

21

21. In this cut-away of the Bosch injection pump, for 4-cylinder engines, the high pressure pump/distributor head assembly can be seen to convert the rotation of the pump shaft and cam plate (**1**) into fore and aft motion of the plunger (**2**). The roller ring (**3**) remains stationary while the cam plate and pump plunger are driven. The cam plate turns against the four rollers (**4**) in the ring, so forcing the plunger to slide rapidly fore and aft, pumping fuel to injection pressure (as high as 700 bar).

Fuel is supplied by the low-pressure vane pump (**7**) to the bore of the hollow plunger via an inlet passage (**6**). With the inlet passage uncovered, the chamber (**8**) charges with fuel. As the open end of this plunger moves back into this chamber, the fuel is pressurised and enters the plunger. As the plunger reaches the limit of its travel, fuel is pumped at high pressure from an outlet port (**9**) to an injector, as an outlet to the relevant port becomes uncovered by a slit in the plunger.

At the opposite end of the pump plunger is a control spool (**10**), which is a sliding fit over the plunger. As the piston reaches the end of its travel, the spool (**10**) uncovers a passage in the plunger, letting the fuel out and relieving the injection pressure. The position of the control spool is set by the operator or vehicle driver via the speed control or accelerator, and also by the action of the governor.

High-pressure outlet passages to which injector pipes are connected are positioned radially around the pump plunger. As the plunger turns, fuel is distributed at high pressure to each one, in a similar manner to a rotor arm distributing a voltage to each segment of the distributor cap.

Injection Timing Mechanism

The injection pump incorporates an automatic timing advance device (***Fig 20, Item 5***) which compensates for the relative delay in injection and ignition as the engine speed increases. This device is controlled by fuel pressure within the injection pump, which increases in direct relation to engine speed, turning the pump's roller ring slightly.

Injection Pump Shut-off

22. Under normal circumstances, a diesel engine is stopped by cutting off its supply of fuel. Solenoid plungers are common but on many engines the action is carried out manually.

With the solenoid in its 'relaxed' state, the plunger is forced by a spring to obstruct

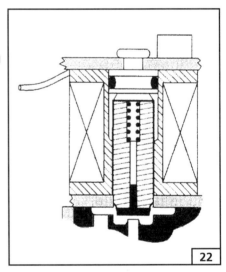

22

a fuel supply passage in the pump. But when the auxiliary switch is turned to ON, the solenoid is actuated by electric current, so pulling the plunger out of the fuel passage, and allowing fuel to enter the high pressure pumping section. However, marine main propulsion engines are never equipped with this type of 'energised to run' solenoid, but are always fitted with a solenoid which is 'energised to stop', for marine fail-safe reasons.

The In-line Type Injection Pump

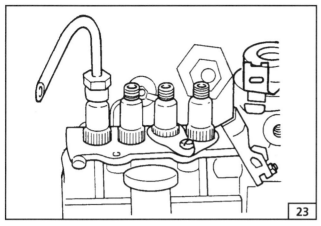

23. The in-line fuel injection pump consists of a series of individual plunger type pumps, one for each cylinder. These pumps are coupled together in a straight line, hence the name. They are actuated by a camshaft, which is in turn driven by the engine. Although the stroke of these pumps is always the same, the length of the effective stroke is varied by means of a helical groove machined into the side of each plunger, and a 'spill port' drilled into the side of each pump body. Whenever the groove and spill port line up, fuel can escape past the plunger, and so fuel is not pumped to an injector until the plunger has travelled past the point where groove and spill port line up. This point can be varied by twisting the plunger, and to achieve this all of the individual plungers are coupled together by a fuel 'rack' which slides backwards and forwards, operated by the speed control, and in doing so twists all of the plungers to the desired position.

Unit Injectors

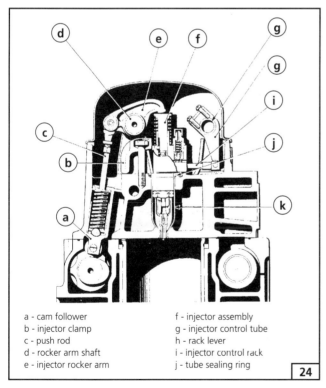

a - cam follower
b - injector clamp
c - push rod
d - rocker arm shaft
e - injector rocker arm
f - injector assembly
g - injector control tube
h - rack lever
i - injector control rack
j - tube sealing ring

24. Unit injectors are individual units mounted in the cylinder head in the positions usually occupied by the injector, (or spark plugs on a petrol engine). Each unit performs the function of both injection pump and injector. They are actuated by the camshaft, via additional cams situated between those operating the inlet and exhaust valves.

25. The 'fuel pump' part of the unit injector is very similar to a single element of the in-line pump. However, there is no rack to control and co-ordinate their operation. Each unit injector is controlled by its own linkage and must be individually set to give the correct calibration and injection timing.

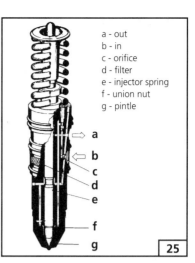

a - out
b - in
c - orifice
d - filter
e - injector spring
f - union nut
g - pintle

Injection Pump Governor

Engine speed is controlled by both the operator and a governor. The governor is needed because the diesel engine is not self regulating, which means that without a governor, if the load on the engine is suddenly increased, the amount of fuel delivered to the engine can go way beyond the amount it can burn. The result is foul black smoke from the exhaust.

For a given engine load a maximum fuel stop can be fitted to the pump to prevent excessive injection, but if the load is removed from the engine (e.g. by a vehicle going downhill or by reducing the power taken from a generator) the simple fuel stop doesn't prevent the engine speed from increasing unchecked, so the engine can literally rev itself to bits! Without a governor, the faster the engine runs, the more fuel the pump delivers and the more the engine accelerates...to the point of self-destruction.

This can be extremely dangerous to anyone in the immediate vicinity, especially with industrial engines, and of course the larger the engine the greater the danger. A great many engines are fitted with overspeed switches designed to stop the engine if its speed climbs too high. These protection devices usually stop the engine independently of the fuel system e.g. an air shut-off valve which cuts off the supply of air needed for combustion.

And at the opposite end of the scale, towards idle speed, fuel output from the pump decreases as engine speed lowers. Without a governor the engine will just stop instead of idling.

The governor automatically controls the flow of fuel to the injectors. Some types, which control delivery at idle and maximum speeds only, are called 'two speed' governors. Other types, which control speeds at the two extremes and at all speeds in-between, are called 'all speed' governors.

There are four categories of governor:
• mechanical governors, which may be either two speed or all speed.
• hydraulic governors, which may be either two speed or all speed
• pneumatic governors, which are all speed.
• electronic governors. which have normally been found where very tight control of engine speed is necessary e.g. generating sets. The increasing use of computer controlled engine management systems means that the use of this type of governor is becoming more common.

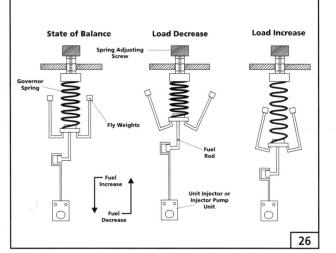

26. A mechanical governor, which is normally built into the fuel pump, has rotating centrifugal weights whose positions are sensitive to engine speed. If engine speed increases above the desired setting, these weights are thrown further outwards. This movement adjusts the position of the fuel controller and reduces the amount of fuel pumped to the injectors. A reduction in speed has the opposite effect.

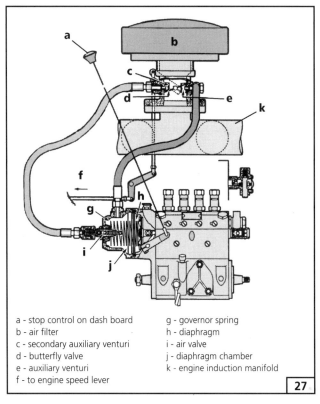

a - stop control on dash board
b - air filter
c - secondary auxiliary venturi
d - butterfly valve
e - auxiliary venturi
f - to engine speed lever
g - governor spring
h - diaphragm
i - air valve
j - diaphragm chamber
k - engine induction manifold

27. Pneumatic governors work on the venturi principle. A venturi control unit fitted in the engine's intake manifold connects by a pipe to a diaphragm unit linked to the pump control rack. The position of the butterfly valve in the venturi, under the control of the driver, determines the depression in the intake manifold. This partial vacuum moves the pump rack via the diaphragm unit. At any fixed setting of the throttle valve an increase in engine speed results in more vacuum which reduces the quantity of fuel injected. Similarly, fuel quantity increases with a fall in speed.

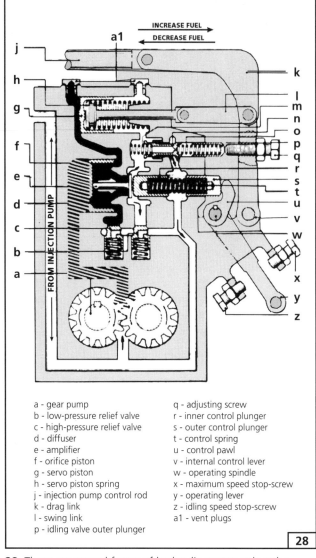

a - gear pump
b - low-pressure relief valve
c - high-pressure relief valve
d - diffuser
e - amplifier
f - orifice piston
g - servo piston
h - servo piston spring
j - injection pump control rod
k - drag link
l - swing link
p - idling valve outer plunger
q - adjusting screw
r - inner control plunger
s - outer control plunger
t - control spring
u - control pawl
v - internal control lever
w - operating spindle
x - maximum speed stop-screw
y - operating lever
z - idling speed stop-screw
a1 - vent plugs

28. There are several forms of hydraulic governor but the essential parts are a gear pump, driven by the fuel pump shaft, which generates a pressure depending on engine speed and the setting of a controllable leak-valve. The resulting pressure, applied to a hydraulic cylinder, operates the pump rack against a balancing spring.

29. Electronic governors measure engine speed via electronic or electro-magnetic sensors. The governor compares the actual with the set speed and through electronic actuators adjusts the flow of fuel to the injectors. Electronic governors have a very rapid response time. This is the AMBAC governor fitted to Perkins CV8 engines.

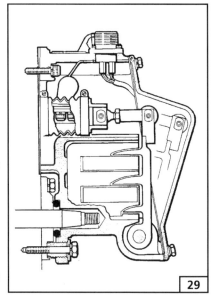

The Injectors

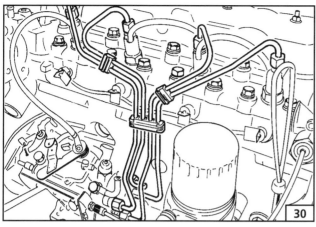

30. The injection pump is worthless without the injectors, which are essential to spray fuel into the engine cylinders. Running from the pump or unit injectors are steel fuel pipes, one per cylinder, which pass metered fuel at high pressure from the pump delivery valve (a sprung outlet valve) to the corresponding injector. Each pipe has a high-pressure screw fitting at each end, very similar to those found on vehicle brake pipes. This is a typical injector and pipe layout for a four cylinder diesel engine.

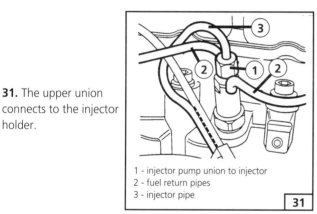

31. The upper union connects to the injector holder.

1 - injector pump union to injector
2 - fuel return pipes
3 - injector pipe

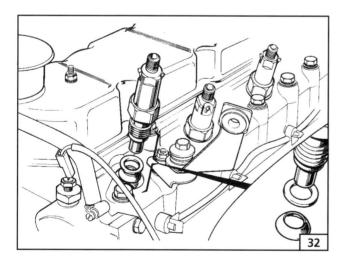

32. This is either screwed or clamped into its bore in the cylinder head with a gas-tight seal between injector and head and this typical indirect injection installation also includes a heat shield. (Illustration, courtesy Ford Europe Ltd)

33. This is a layout for a direct injection installation and shows the low-pressure fuel return or leak-off pipe (**2**).

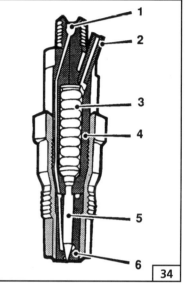

1 - high pressure pipe union
2 - banjo bolt - leak-off pipe
3 - clamp bolt assembly
4 - pedestal
5 - injector
6 - seat washer
7 - locating ring

34. Fuel enters the injector at the inlet (**1**) and passes down an internal drilling in the body (**4**) to a chamber in the nozzle (**6**). The holder combination comprises the holder and injector nozzle (**6**), and contains a compression spring (**3**) which acts on a pressure spindle which incorporates a nozzle needle at its lower end (**5**). (Illustration courtesy V L Churchill/LDV Limited).

The injectors return excess fuel to a point 'upstream' of the fuel injection pump, usually the fuel tank or the fuel filter via low-pressure piping (**2**).

35. The injector holder secures the injector nozzle in the cylinder head and seals the combustion chamber. High-pressure fuel from the injection pump enters the top of the injector and passes to a chamber around the conical face of the nozzle needle. The fuel

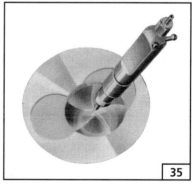

pressure in this chamber acts upon the conical face, so forcing the nozzle needle upwards against spring pressure. This causes the nozzle to open, and fuel to be injected into the combustion chamber. As injection comes to a close and pressure falls off, the spring returns the needle to its seat and injection is completed. A certain amount of fuel by-passes the spindle and exits at low pressure into the fuel return line mentioned earlier. This back-leakage of fuel is vital, since it provides lubrication for the spindle, which can reciprocate in the nozzle bore many thousands of times a minute.

That's the basic operation of the injector, but it's worth adding that there are different types of injector with varying characteristics determined by the design of the combustion chamber with which they are used. The formation of the fuel injection spray is greatly affected by the length, shape and size of the injector nozzle. If the opening is short, fine atomisation is obtained, but with a long opening the spray has more penetration; a compromise invariably has to be made between these two situations.

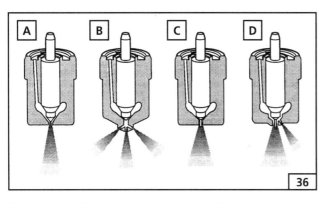

36. In general though, there are two basic injector types: the orifice type for direct-injection engines, and the throttling pintle type for indirect-injection engines. Common types of injector include the single-hole (**A**), multi-hole (**B**), pintle (**C**), delay and Pintaux types (**D**), each of which is suited to a different design of combustion chamber. The last three are generally intended for indirect-injection applications. (Illustration, courtesy Dieseltune)

With the pintle type a small cone extension on the end of the needle produces a pre-injection. As it opens, the nozzle needle initially permits the passage of only a little fuel through a very narrow annular gap (throttling effect). As it opens further, due to the rise in fuel pressure, the cross-section of the flow increases, and the main proportion of the fuel is not injected until towards the end of the needle lift. Combustion and engine operation become smoother with the throttling-pintle nozzle, because cylinder pressure increases more slowly. The pintle nozzle is the common type fitted to indirect-injection cars.

DI and IDI - OK?

In case you remain unsure of the differences between DI and IDI, allow us to explain further.

Both injection systems are used in diesel engines of current manufacture, IDI being usually employed for passenger car diesels, and DI for most other applications. The fundamental differences between the two types lies in the design of the combustion chambers.

• The direct injection type has a combustion chamber formed in the top of the piston and the injector injects fuel into the combustion chamber.
• The indirect type, however, injects into a separate (smaller) chamber in the cylinder head, known as the pre-combustion chamber. This communicates with the cylinder via a narrow passage. The top of the piston has either a small depression to direct combustion or is flat. The injector is positioned to inject into the pre-combustion chamber.

Research work on the diesel engine in the 1930s led to a better understanding of the combustion process, and by the end of that decade these two fundamental types of combustion chamber were emerging from the proliferation of available designs. British engineers concentrated on improving specific fuel consumption, by increasing thermal efficiency through the use of direct injection combustion chambers. Continental designers, however, were more concerned with achieving the smooth and progressive burning of fuel by a system of pre-combustion using an 'ante-chamber' to originate a controlled flame, which then spread to the main combustion chamber.

Indirect-injection

In the ante-chamber type of cylinder head, when the piston rises on the compression stroke, air is forced into a chamber connected to the head by one or more passages. This chamber is maintained at a high temperature and a certain degree of air turbulence is created by the airflow into it. The fuel injector is positioned in the ante-chamber and ignition of the fuel spray creates rapid expansion which forces the remaining unburned fuel particles out of the ante-chamber into the main chamber where burning continues.

The combustion process in the ante-chamber engine was designed to give a more sustained 'push' to the piston than would be the case if the burning of the charge took place in the open cylinder, for it was assumed that ignition in the diesel engine was more rapid than in the petrol engine. This was only true to a point; in the diesel engine there is a time lag between the commencement of fuel injection and the initiation of combustion. This was not clearly understood in the pioneering days of combustion chamber design, so the pre-combustion chamber was an attempt to control the combustion process.

Stopping Smoking

After further years of work, it became possible to obtain controlled and complete combustion in an open chamber by means of more homogeneous mixing of fuel and air through better controlled turbulence, more refined injection equipment and improved fuel quality. It was found that smoke emissions from the exhaust were invariably caused by even the smallest unburned fuel droplets, and the perfect matching of fuel spray pattern to combustion chamber shape and turbulence characteristics became an essential requirement for the avoidance of this phenomenon.

37. The greatest advance in combustion chamber design was made by Sir Harry Ricardo was made with the innovative Comet IDI engines. The Comet engine developed the idea of a sort of 'ante-chamber', separate from the combustion chamber and joined by a very narrow passageway.

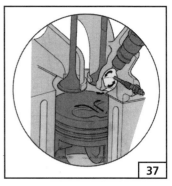

As the piston rises, air in the combustion chamber and in the pre-combustion chamber is heated, as with all diesel engines. Just before top dead centre (TDC), fuel is injected into the pre-combustion chamber. The mixture ignites, and projects itself across the top of the piston (which is almost at TDC). In the case of the Comet engine, there is a pair of shallow depressions in the piston crowns. The already ignited mixture meets the heated, compressed air above the pistons, and the depressions help to create a pair of swirling vortices, as the combustion process is completed.

The Comet engine lead the way towards today's modern designs of IDI engines in providing greater engine flexibility and less smoke than any previous IDI designs.

Direct Injection

38. Although in the open chamber of a direct injection engine, compression space can be reduced to give any desired compression ratio (the limiting factor being the valve to piston clearance), a high compression ratio is not necessarily required for ignition, nor even for cold starting. But, in its simplest form, the open

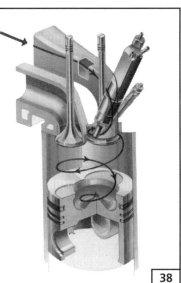

chamber does not give the required turbulence characteristics for the prevention of smoke emission. It does, however, have the advantage of diluting the fresh air charge less with residual gases.

Thus the direct injection engine is efficient in most respects: it is a reliable cold starter, returns good specific fuel consumption because of its superior thermal efficiency (typically 25% better than the IDI engine) but it is nonetheless inclined to roughness in operation and incomplete combustion. Its significantly better fuel economy is one of the main reasons for its popularity where operational refinement is not of paramount importance.

IDI allows smoother, more progressive burning of fuel and so quieter, more refined and often cleaner engine running - hence its common application in passenger cars. But it returns comparatively poor economy because the addition of the pre-combustion chamber substantially increases the surface area of the total combustion zone, which means that more useful combustion heat is dissipated to the cylinder head than with DI designs. Operational efficiency is also reduced by the pumping losses incurred by the air-charge passing in and out of the pre-combustion chamber via a narrow passage.

Recent design advances, particularly in the fields of fuel injection, turbocharging and intercooling, are narrowing the practical differences between DI and IDI.

The Fuel Filter

39. The internal workings of the injection pump and injectors are so finely machined, that the smallest of dirt particles will rapidly wreck these expensive components. An injection pump can be turned into scrap by just 5 grams of microscopic dirt!

Diesel fuel acts as both lubricant and coolant for the moving parts of the injectors and injection pump, so just as it is vital to keep engine oil filtered, so it is even more essential to filter the fuel that passes through the fuel injection system

Every diesel engine has a fuel filter connected in the pipework upstream of the fuel injection pump, usually mounted on the engine, but sometimes away from it. The filter consists of a pleated roll of special paper capable of trapping particles of 5 to 20 microns in size.

How Does It Work?

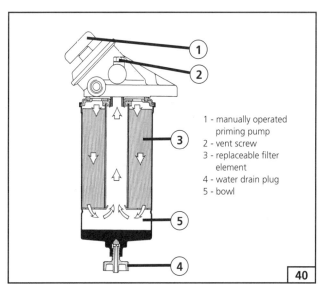

1 - manually operated priming pump
2 - vent screw
3 - replaceable filter element
4 - water drain plug
5 - bowl

40. Fuel enters the filter element via the filter head or cap and flows around the pleated paper element, which both filters the particles and separates any residual water from the fuel. The fuel passes through the element and up the centre tube to return to the filter head. Filtered fuel then passes to the fuel injection pump. Any water is allowed to collect at the bottom of the filter where it can be drained off.

The filter element will eventually become clogged with dirt if it is not renewed at the specified service intervals. A clogged fuel filter will mean an engine breakdown either due to the fuel flow being reduced to the point where the engine will not function properly, or due to injection pump damage by dirt particles which have bypassed the filter element.

Removing Water

41. Water is very damaging to the fuel injection system, since it is both corrosive and a non-lubricant. There is a great danger of water contamination in marine environments, but water can be found in diesel fuel if only because it's impossible to avoid condensation in a fuel tank. So the fuel system is designed to separate water

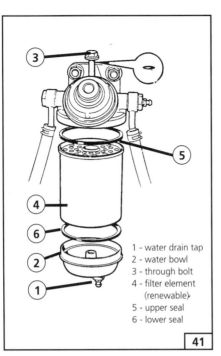

1 - water drain tap
2 - water bowl
3 - through bolt
4 - filter element (renewable)
5 - upper seal
6 - lower seal

41

from the fuel and trap it for later removal. This is easy since water is heavier than diesel fuel, and so will naturally sink to the bottom of the filter assembly.

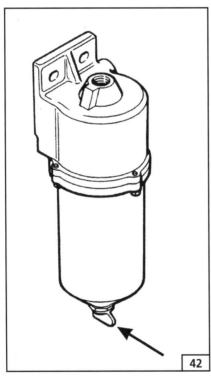

42. Water should be drained from separators and filter bowls at least as often as the specified service intervals. This filter is fitted with a drain thumb-screw (arrowed).

42

Heating The Fuel

Why would you need to heat diesel fuel? Because it has a high wax content which crystallises out at very low temperatures, and can clog the fuel filter. The result is a misfiring, stalling or even a non-starting engine. The application of a little heat soon gets things on the move again though.

43. Although modern winter grade diesel fuel has additives to prevent waxing down to temperatures as low as -20 degrees C, many diesel engines are fitted with a fuel heater mounted either within or around the fuel filter or in the fuel supply line. Heaters, which can be either of the

43

electric or of the coolant type, can be switched on or off automatically by means of a thermostat.

An electric fuel heater can usually be recognised at a glance by the presence of electrical wiring to the heater unit, but coolant-type heat exchangers (usually bolted to the cylinder head) are less obvious.

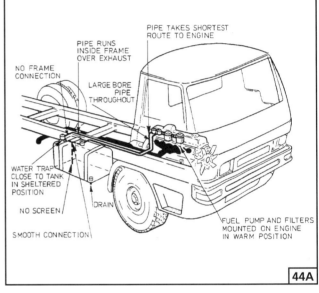

44A

44A. A well designed fuel system will help to overcome some of the problems of fuel waxing...

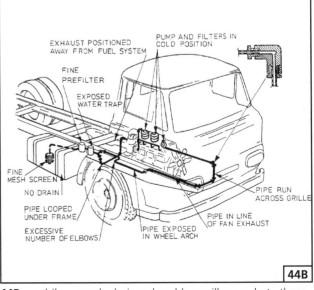

44B

44B. ...while a poorly designed problem will exacerbate them - and introduce other problems, too.

The Pre-heating System

Although the compressed intake air gets very hot right from the first stroke, the cold cylinder walls conduct heat away, so the diesel engine benefits from extra heat when starting from cold.

45. The most familiar of these pre-heating devices is the 'glow plug' which is fitted to most diesel engined cars. The glow plug is a long thin electrical element which protrudes into the combustion chamber (pre-combustion chamber on IDI engines).

When the operator (or driver) turns the auxiliary key to the pre-heat position, the glow plugs heat up to about 800-900 degrees C. After a few seconds of pre-heat the engine can be started; the whole process takes from two to 20 seconds, depending on the engine.

The 'port-heater' operates in a manner somewhat similar to the glow plug. This is an electrically heated coil that heats the intake air as it enters the induction manifold and is usually used in a group of two or three.

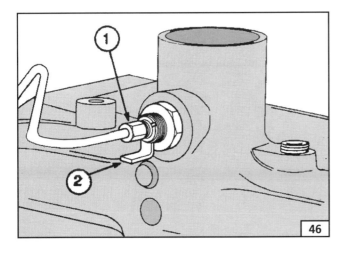

46. The fuelled starting aid is an electrically operated device (with an electrical contact - **2**), which ignites a controlled amount of diesel fuel (supplied by a fuel pipe - **1**) in the induction manifold. A heater coil in the body expands a valve holder to allow fuel to flow into the device, where it is ignited by the hot coil. The heated air is drawn into the combustion chamber when the starter motor turns the engine over.

Excess Fuel

Often a fuel pump will be found to have an 'excess fuel' device, which temporarily increases the amount of fuel injected during starting, and is comparable to a petrol engine's 'choke'. However, remember that it is not a choke, and must always be used as instructed in the User's Handbook.

Turbocharging

47. Diesel engines are turbocharged not only in order to increase their performance, but also to give smoother running and decrease smoke emissions. By turbocharging an engine you pack more air into the cylinders which allows more fuel to be burned, and so more power to be produced.

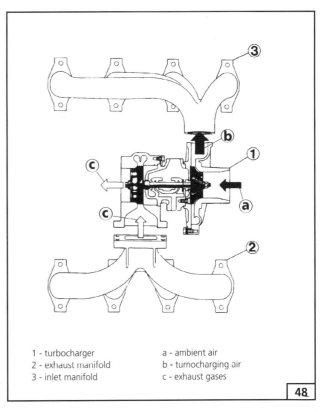

1 - turbocharger
2 - exhaust manifold
3 - inlet manifold
a - ambient air
b - turnocharging air
c - exhaust gases

48. About 40% of the useful heat energy generated by an unturbocharged engine is lost in the exhaust gases. However, the turbocharger is driven by some of this otherwise wasted energy, which results in a more efficient engine.

Intercooling

Intercooling is commonly applied to turbocharged diesels in order to decrease the temperature of the intake air. Since the turbocharger, in the act of compressing the air, heats it up and so makes it expand, cooling it will increase the amount of air in a given volume, so packing yet more combustion air into the cylinders. One of three types of intercooler may be found:

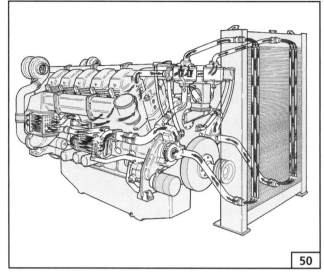

● **49.** Air-cooled intercooling, where the compressed air is passed through a radiator matrix similar to the familiar one used for vehicle engine cooling.

● **50.** Jacket-water cooled intercooling, where the compressed air is passed through a heat exchanger where engine coolant absorbs some of the heat.

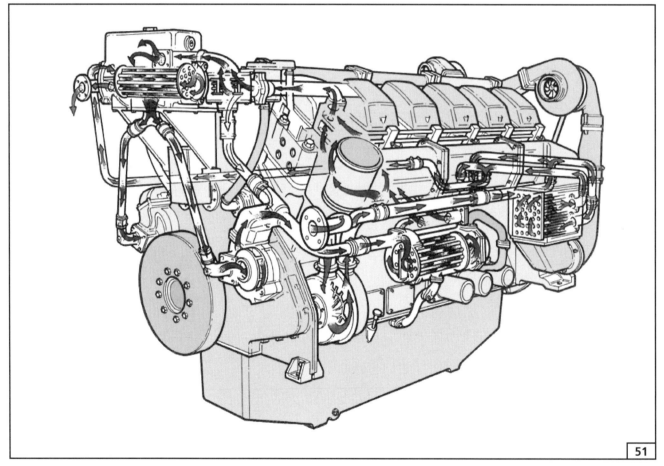

● **51.** Raw-water intercooling, similar to jacket water intercooling except that the compressed air is cooled by a totally separate source of water such as that from a river, the sea, lagoon or similar. The use of this type of intercooling is of course restricted to either marine or stationary engine applications.

Before carrying out any of the checks or work detailed in this Chapter please read carefully CHAPTER 1, SAFETY FIRST!

CHAPTER 3 · STARTING PROBLEMS · PART A

CHAPTER 3
STARTING PROBLEMS

Although experienced diesel mechanics may often be able to go straight to the cause of a problem, the best advice to anybody else is - don't leap to assumptions! There are many reasons why an engine might not start. In the long term, you will save time and effort by following the logical sequence of checks listed here and detailed later in the chapter. A systematic elimination of each check by testing rather than by hunch, before moving on to the next check, will increase your chances of finding that fault.

Remember that the great majority of failures are caused by lack of fuel at the injectors, or by a battery that is too weak to supply your diesel engine with a decent cranking speed. So, before assuming you have a problem, please do physically check that you actually have fuel in the tank, any tank shut-off valves are open, the battery is in good condition and, if a battery master or stop/run switch is fitted, it is in the ON or RUN position.

How To Use This Chapter

This Chapter is divided into three parts:
• PART A lists possible faults and remedies.
• PART B shows you how to check for each fault.
• PART C deals with how to carry out necessary repairs.

Each fault has the same 'Check' number in PART A and PART B, and the same number (now a 'Job' number) in PART C.

Chapter Contents

PART A: FAULT FINDING CHECKLISTS

SAFETY FIRST!

• *Industrial engines often have emergency stop devices which prevent the engine being restarted until the cause of the shut-down has been corrected.*

• *You should always investigate and correct the cause of any automatic shut-down as there is the possibility of damage or injury if the engine is restarted.*

• *Many industrial engines can also be started remotely and it is imperative that a check is made to ensure the engine is safe to start.*

IMPORTANT NOTE: Most engines you come across will be electric-start but, occasionally, some will be found to have compressed air starters and other small hand-cranked engines do exist. There is really very little that can be done if a compressed air starter fails. If you have a hand-cranked engine it will be immediately apparent whether the engine has seized. In both cases the comments relating to the electric starting system will not apply.

FAULT 1:
ENGINE WON'T TURN

	CAUSE	REMEDY
Check 1.	❑ Engine seized.	❑ Various see *Part B*.
Check 2.	❑ Battery partially or fully discharged.	❑ Charge or replace battery.
Check 3.	❑ Starter motor jammed.	❑ Free or replace starter motor.
Check 4.	❑ Starter switch or circuit wiring faulty.	❑ Repair or replace wiring or switches.
Check 5.	❑ Starter motor solenoid faulty.	❑ Repair or replace solenoid.
Check 6.	❑ Starter motor faulty.	❑ Repair or replace starter motor.

FAULT 2: ENGINE TURNS BUT WON'T START

Diesel engines do not have the problems associated with an electric ignition system. However, there are still many reasons why a diesel engine will fail to start. At its simplest, a diesel needs fuel, air, sufficient cylinder compression and a high enough cranking speed before it will start. Many factors can be present, either singly or in combination, to prevent either one or all of the essential requirements doing its job.

	CAUSE	REMEDY
Check 7.	❑ Slow cranking speed.	❑ Charge or replace battery. ❑ Ensure electrical connections are good. ❑ Remove load from engine. ❑ Make sure oil is correct grade. ❑ Open decompressor.
Check 8.	❑ No fuel at injectors.	❑ Put fuel in tank. ❑ Open fuel cock. ❑ Check cut-off solenoid. ❑ Bleed air from system. ❑ Repair or replace lift pump. ❑ Eliminate fuel waxing.
	CAUSE	REMEDY
		❑ Repair or replace fuel injection pump. ❑ Remove blockage from fuel lines or injectors.
Check 9.	❑ Water in fuel.	❑ Drain water from system.
Check 10.	❑ Low compression.	❑ Close decompressor. ❑ Overhaul engine. ❑ Check air filters. ❑ Check valve clearances.
Check 11.	❑ Insufficient air flow.	❑ Clean or replace air filter. ❑ Remove obstruction from induction system. ❑ Replace hose. ❑ Open emergency air valve.
Check 12.	❑ Cold start aids not working.	❑ Repair or replace cold start aids.
Check 13.	❑ Blockage or restriction in the exhaust system	❑ Clear blockage.

PART B: FAULT FINDING STEP-BY-STEP

FAULT 1: ENGINE WON'T TURN

Check 1. Engine seized.

IMPORTANT NOTE: Continued attempts to start an engine that has seized can result in permanent and serious damage to the engine. A seized engine should be stripped-down and any damaged components replaced.

making it easy / • *If the engine's electrical system contains an 'ignition' warning light, which will normally go out when the engine starts, it can be a useful guide to the reason for the fault. If the light does not come on, the battery is either disconnected, discharged or there is a break in the electrical circuit. A bright light that goes out when the start switch is activated indicates that the battery condition is low, the battery connections are bad, or the engine or starter motor has seized. If the light stays bright, it means that the starter solenoid is faulty or the starter motor brushes are worn.*

A seized engine is unusual. However:

• If the engine has not been used for a very long time, without proper protection, it is possible for corrosion and deposits to cause the pistons to jam in the bores (or the valves to stick in their guides).

• An engine that has been in regular use may seize because of overheating, which causes piston seizure; or lack of lubrication, which causes bearing seizure.

• Broken or damaged valves and springs; foreign objects such as a nut or washer in the cylinder bore; fuel, oil or water in the bore, causing a hydraulic lock, all stop the engine from turning over.

• A broken camshaft drive belt or incorrectly assembled camshaft drive assembly can also be a cause of seizure by allowing pistons to strike and damage valves.

1A. Check whether the engine is seized by using a suitably sized socket or ring spanner on the crankshaft pulley nut to slowly turn the engine. If the engine cannot be turned in either direction, the cause is probably due to piston or bearing seizure. If it can only be turned against the normal direction of engine rotation the cause is probably due to one of the other reasons shown here. The only sensible solution is an engine strip-down and overhaul.

INSIDE INFORMATION: If you have an engine that only needs to be used infrequently, you should run it at regular intervals, long enough for normal operating temperature to be reached, with oil changes carried out at the calendar periods specified by the manufacturer. Alternatively, if the engine is to be stored, it should be protected against corrosion damage, in accordance with the manufacturer's instructions.

The source of any foreign objects should be investigated and rectified before a new or rebuilt engine is refitted and used. Similarly, the reasons for overheating or the cause of any oil, fuel or water ingress should be remedied before a new or rebuilt engine is fitted.

FACT FILE: HYDRAULIC LOCK CAUSES

A 'hydraulic lock' in one or more cylinders will stop the engine from turning and gives the same symptoms as a seized engine.

Causes can be:

• Leaking thermostart (cold start aid)
• Blown head gasket (coolant in cylinder)
• Bad design of exhaust system (marine)
• Bad design of inlet system (early Montego car engine, for instance)
• Porous water-cooled exhaust manifold
• Porous water-cooled inlet manifold
• Leaking injectors

Check 2. Battery discharged.

SAFETY FIRST!

• The gas given off by a battery is highly explosive. Never smoke, use a naked flame or allow a spark to occur in the battery compartment. Never disconnect the battery (it can cause sparking) with the battery caps removed.

• Batteries contain sulphuric acid. If the acid comes into contact with skin or eyes, wash immediately with copious amounts of cold water and seek medical advice.

• Do not check the battery levels within half an hour of the battery being charged with a battery charger. The addition of fresh water could then cause the highly acid and corrosive electrolyte to flood out of the battery.

Batteries gradually lose their charge and a battery fitted to an engine that has not been used for some time probably won't have enough energy to turn the engine, especially if the weather is cold. A discharged battery on an engine that is used regularly suggests a faulty charging system.

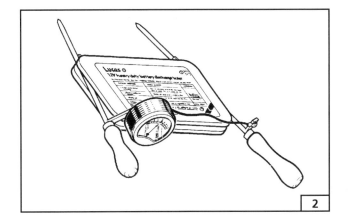

2. IMPORTANT NOTE: A diesel engine's starter motor draws a relatively high current from a battery and using a voltmeter or multi-meter across a battery's terminals does not give a true indication of the battery's ability to deliver the necessary current. Even connecting a headlight bulb across a battery can give a false indication of its condition. Use a hydrometer to check the condition of the electrolyte in each cell or a high current drain discharge tester, such as the Lucas tester shown here.

Charge the battery, replace it with one that is fully charged or jump start it from another battery. A jump start is only recommended when the donor battery is connected to another engine which can be run at the same time. If the engine is in a vehicle, a bump start may get the engine going although this will not be possible if automatic transmission is fitted. If a battery has been discharged for a long time, it will probably have suffered irrecoverable internal deterioration and will not hold a good charge, even after prolonged charging.

Investigate the reason for the battery being discharged. Remove a battery which is not going to be used for a lengthy period and keep it fully charged using a battery charger. If charging in-situ, take note of any warnings contained in the engine or equipment maker's User's Handbook. Some parts of an engine's generating circuit can be damaged by high charging currents.

If a 'good' battery fails to charge when the engine is in use, check out the condition of the charging circuit and rectify any faults.

Check 3. Jammed starter motor.

A jammed starter motor has some of the same symptoms as a seized engine. The engine does not turn, but the 'ignition' warning light will dim - though this can also be the symptom of a partially discharged battery. A jammed starter is normally due to wear on the flywheel ring gear teeth and starter motor pinion teeth but could be a loose motor attachment.

• Check that the starter motor mountings are tight and that it is mounted 'square' on its mounting point.

• Attempt to free the starter motor by rotating the starter motor armature (there may be a square shaft on the back of the motor), or by rotating the engine itself.

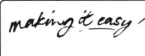

making it easy • If the engine is in a vehicle, put the vehicle in gear and rock it backwards and forwards.

• If the starter remains jammed, loose the motor on its mountings and move it to one side, freeing the pinion.

• If the problem persists remove the starter motor and check that the pinion is free on its shaft. If it has been oiled or greased, it can retain deposits which cause it to stick. Clean it, but do not lubricate it.

If necessary, replace the starter motor pinion or the starter motor.

i INSIDE INFORMATION: The old-fashioned inertia type of starter motor (see **Check 6**) always engages the ring gear in the same places. If no replacement ring gear is available, it may be possible, with careful use of a cold chisel, to drift the ring gear off the flywheel, and refit it, turned by 45 degrees around the flywheel. On a few engines, it may be possible to rotate the flywheel to a new position or replace the ring gear. *i*

Check 4.
Starter circuit wiring faulty.

Starter motors draw a relatively high current from the battery - in the order of 400 to 500 amperes. For this reason, thick cables are used and the supply switching is remotely controlled by a solenoid. The starter switch only has to cope with the low current to activate the solenoid.

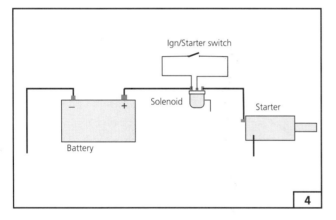

4. Check to make sure that the high current cables and connections are in good condition. On some installations, the engine will be connected to the battery through a common earth (ground) system, like that used on most vehicles, while on others the engine may be connected directly to the battery by a thick cable. If these appear okay, check the condition of the low current wires connecting the battery, starter switch and solenoid.

i INSIDE INFORMATION: If, in an emergency, and if the starter switch becomes faulty, the appropriate terminals can usually be shorted out or by-passed completely. *i*

Check 5.
Starter motor solenoid faulty.

Solenoids are generally reliable and have a long life. However, the external contacts can become dirty and corroded and the internal contacts can sometimes fail because of a build up of deposits.

Listen to or feel the solenoid when the starter switch is moved to the START position. A click or thump indicates that the solenoid is being activated, but only an electrical check will prove whether or not it is doing its job.

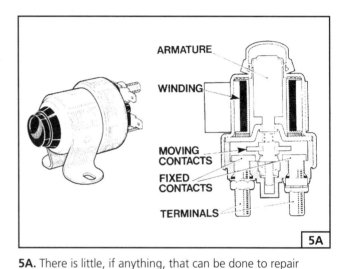

5A. There is little, if anything, that can be done to repair solenoids and the only really practical solution is replacement.

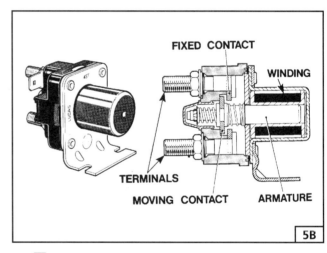

5B. *i* INSIDE INFORMATION: In an emergency the solenoid can be by-passed either by shorting out the contacts or supplying the starter motor direct from the battery. Be sure to use cable at least as thick as that used for the existing connections. *i*

Check 6. Starter motor faulty

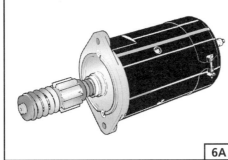

6A

6A. There are two types of starter motor in common use, the inertia (or Bendix) drive...

6B. ...and the pre-engaged type. The pre-engaged type is identified by a solenoid mounted on the body, while the inertia type will have a separate solenoid.

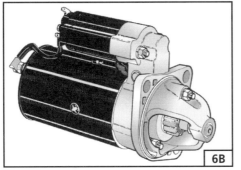

6B

Invariably, a faulty starter motor has to be replaced as there is little the user can do in the way of repair work - unless the problem is due to worn-out brushes. Brushes can generally be replaced quite easily but it is often not that much more expensive to obtain a reconditioned unit.

FAULT 2: ENGINE TURNS BUT WON'T START

making it easy
- Air inlet additives are available which, the manufacturers claim, can assist reluctant diesels into life.
- *They generally contain ether, which is highly flammable, so always carefully read and follow the instructions on the container.*
- *You should also check to make sure the engine manufacturer does not prohibit their use.*
- *If an engine needs one of these 'remedies-in-a-can' on more than the rare occasion, there is a fault which needs to be put right.*

Check 7. Slow cranking speed.

A very low cranking speed, such as the engine only just turning, is immediately obvious. At higher speeds, the only certain way to check whether the cranking speed is down is to measuring engine speed with a tachometer. The electrical type uses a light source and a reflective spot on the shaft, while the mechanical type has a small wheel that is pressed against the shaft.

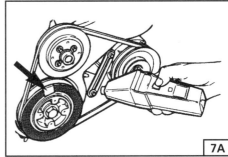

7A

7A. An electrical diesel-engine tachometer requires a little preparation. First, a piece of adhesive reflective tape is applied to the crankshaft pulley (arrowed). It is important that this tape is applied only to either the crankshaft pulley or some other component that rotates at crankshaft speed. The alternator, water pump and others such as bilge pump pulleys will not rotate at crankshaft speed and will give a false reading. It is then simply a matter of pointing the tachometer at the pulley, in a position where the reflective tape will pass through the light beam of the tachometer. The engine should now be rotated on the starter motor, and the engine speed will be displayed on the tachometer display. A mechanical tachometer needs to be held against the crankshaft pulley as it rotates.

Make a note of the cranking speed and compare this with the figures given in the manufacturer's manual.

Use a hydrometer to check the battery charge level. If the readings are low, recharge the battery. See **Chapter 3, Part C, Job 2**.

Carry out the checks in **Chapter 3, Part C, Job 4, Step 3**.

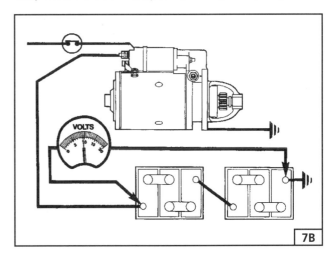

VOLTS

7B

7B. Check the battery voltage under load. Connect a voltmeter across the battery terminals, and operate the starter motor. The voltmeter will give a reading which will vary with the engine, the battery, and the starter motor type. Once again, consult your engine's Workshop Manual for specific detail. If the voltage read-out is low, then the starter motor should be removed and replaced or sent to a specialist for bench testing.

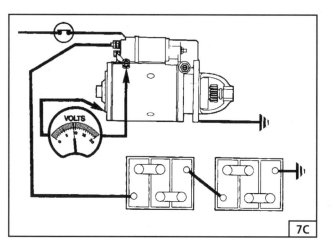

7C. Check the voltage under load at the starter motor. Connect a voltmeter between the starter motor terminal, and earth/ground. When the starter motor actuating switch is operated, the voltage should be within the limits specified by the manufacturer. If the voltage is normal, then renew or repair the starter motor. If the voltage reading is too low, then this indicates a fault in either the wiring or electrical components fitted in the circuit between starter motor and battery.

Unless you are completely unfamiliar with the engine, or the weather is extremely cold, it is unlikely that the wrong grade of oil is causing the engine to turn slowly. The only sure way of checking the oil's viscosity is to take a sample and have it tested. On all but the largest engines it may be cheaper and more convenient to simply drain and replace the oil with the correct grade.

Check 8. No fuel at injectors.

Carry out the obvious checks first, such as making sure there is sufficient fuel in the tank, that the liquid in the tank is actually fuel, that any fuel stop cocks are open and any run/stop switches are in the RUN position.

Check the operation of the fuel cut-off solenoid. Feel or listen to the solenoid while operating the starter switch, it should be possible to sense the movement of the solenoid.

Remove the connectors and measure the voltage between the solenoid supply and earth/ground. For 'energised to run' solenoids full battery voltage should show on the meter when the starter switch is on. 'Energised to stop' solenoids should show full voltage in when the starter switch is off.

Use an Ohmmeter to check the resistance of the solenoid coils, the correct value will vary depending on solenoid type and manufacturer. Generally, zero resistance indicates a short circuit and very high resistance indicates an open circuit.

If the electrical tests show a problem or you have doubts about the solenoid the only practical solution is replacement.

Check the fuel supply system or air leaks. Remember the pumps suck fuel so if there are any leaks they'll also suck air in. Make sure all the connectors are tight and examine the delivery pipes for damage or deterioration. Some air leaks can be very difficult to detect.

If air in the fuel system is suspected, bleeding is necessary. Undo the inlet to the high pressure fuel pump or, if fitted the inlet of the lift pump. When you are sure that only fuel is flowing and any air has been completely removed remake the connection.

8. If the lift pump has a hand primer, disconnect the inlet to the fuel injection pump and hand prime the pump until only fuel flows from the pipe. Before reconnecting the pipe to the injection pump crank the engine for a few seconds. If there is no output, suspect a faulty lift pump. It is possible to repair some pumps and you should contact the supplier to see what repair kits are available. Otherwise the pump will have to be replaced.

Check the engine or fuel injection pump manufacturer's instructions to see whether the pump is self-priming. Follow any instructions for bleeding but do not randomly undo bolts or screws on the pump in a vain search for a bleed valve. Catastrophic pump damage can be caused by undoing the wrong screws.

If the weather is very cold, below freezing, fuel waxing may be the problem especially if the tank was last filled in the summer. Check the condition of the fuel at a convenient point in the system and if you suspect waxing try warming the fuel delivery system.

> **SAFETY FIRST!**
>
> • *Although much safer than petrol, diesel fuel will ignite if heated sufficiently, especially if heated with a naked flame. Heat fuel with caution.*

It is possible to check whether fuel is reaching the injectors by removing them and checking for spray when the engine is cranked. Remove one injector at a time and direct the spray into a container.

No spray, or only a dribble from the injectors indicates a general problem with the fuel injection pump as it is unlikely that all the injectors would be blocked. A spray from some of the injectors could indicate a fuel pump problem or a blockage. Change the injectors round, if there is no change the problem is with the pump or high pressure fuel line.

Check 9. Water in fuel.

Water and fuel don't mix, so it is usually quite easy to separate water from the system. A good fuel system will have several locations where water can be drained but there should be at least one drain point. This will probably be at the fuel filter. If there is a lot of water in the system and the engine has been idle for some time, vital components may be corroded. Particles of corrosion are very abrasive and they may cause wear and premature failure of fuel injection components.

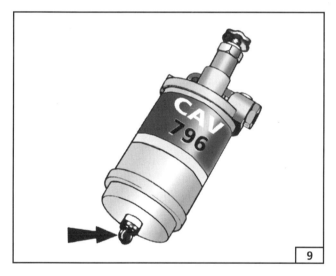

9. Drain water from the system wherever a drain tap is fitted. Crank the engine and repeat the process. Thereafter, it may be necessary to crank the engine for a time for any remaining water to clear.

making it easy • *Water in the fuel is normally due to condensation in the tank, although contaminated fuel and water ingress during refuelling are not unknown.*

• *Keep the tank full, especially if the engine will not be used for long periods.*

• *In humid conditions when there is a big temperature difference between night and day, it can pay to top the tank up at the end of the day.*

Check 10. Low compression.

• If the cylinder compressions are too low, the air cannot be compressed to the temperature needed to ignite the fuel.

• An engine with low compressions may turn over at a good speed and show no obvious faults. It may start without difficulty on a hot summers day but be impossible to start in the middle of winter.

• Piston ring and cylinder bore wear, or valves that are not closing or seating properly are the most common causes of low compressions.

• A blocked air filter can also be a cause.

• However, if the engine is fitted with a decompressor, check that you are not leaving it open once the engine is up to a good cranking speed.

• Also, check that the valve clearances are correct and that all valves close fully.

A cylinder compression test using a tester designed for diesel engines will give some indication of the problem. Do all cylinders have low compression or just a few? Low compression can be due to poor valve seating, blowing head gaskets, cylinder bore wear and piston ring wear. All these causes will need a degree of strip down to discover the cause and rectify the problem.

10. Piston ring and cylinder bore wear is normally accompanied by high lubricating oil consumption and blue smoke out of the exhaust.

Only a compression test can confirm for certain that low compression is causing the problem - see *PART C, Job 10*. Unfortunately though, identifying the exact cause and curing the problem invariably means a strip-down.

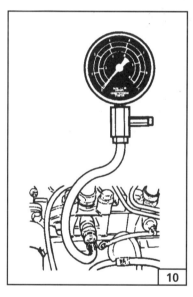

Check 11. Insufficient air flow.

Some industrial engines have an emergency air flow shut-off valve. Check to see if there is one fitted and, if it has been activated, find out why before attempting to start the engine. The most common cause of reduced airflow is simply a blocked air filter. However, if a new one is fitted and the problem persists, check the intake ducting for both damage and blockages. Repair or replace any damaged ducts or hoses.

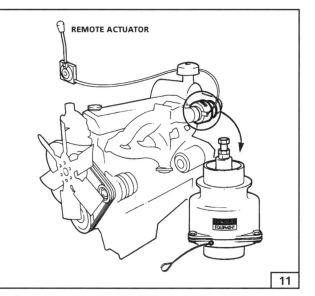

REMOTE ACTUATOR

11. Check for an emergency air shut-off valve. Do not attempt to start an engine if the valve has been activated: find out why and correct the problem first.

Change the air filter at the manufacture's recommended interval or more frequently if the operating conditions are very dusty.

In an emergency, run the engine without a filter but be aware of the damage and rapid engine wear that contaminated intake air can cause.

Check 12.
Cold start aids not working.

The most common type of cold start aid is the glow plug, an essential aid for indirect injection engines. Both the heating element and the duration of operation are electrically operated. As a result, problems may be due to the electricity supply as much as to the glow plugs themselves.

IMPORTANT NOTE: Although subjected to a very hostile environment inside the engine, glow plugs are quite fragile and can be easily damaged when removed. It is unlikely that all glow plugs in an engine will fail at the same time so, if the engine shows no inclination to start at all, the problem is probably with the common supply. An engine that fires on a few cylinders before 'catching' completely may be symptomatic of some failed glow plugs.

i INSIDE INFORMATION: Beware of glow plugs that operate on a pulsed supply and have a low resistance. Connect them to a continuous supply, such as from the battery and they will burn out within seconds. *i*

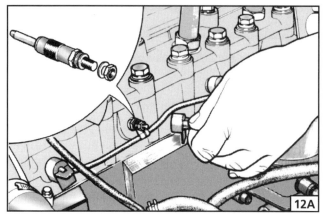

12A. Glow plugs are wired in parallel and power is supplied through a relay (solenoid switch). The location and type of relay will depend on the engine manufacturer and a wiring diagram for the particular engine should be used to identify and locate the relay and associated wiring. Power supply to the relay, timer and plugs should all be checked. Bear in mind that the glow plugs may only operate for 20 seconds, so any checks will have to be quick. (Illustration, courtesy Ford Europe)

There is not much you can do about a faulty glow plug, relay or timer. Replacement with serviceable components is the most sensible course of action.

Some cold starting aids burn fuel in the inlet manifold. These have both an electric and a fuel supply so both areas will need to be checked.

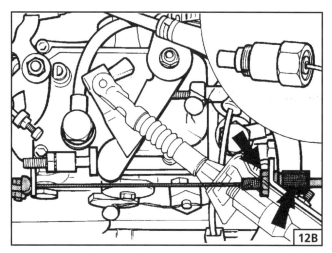

12B. If you have an engine with a cold engine fast idle device, check that the device (see inset) operates and that the cable is correctly adjusted (arrows) at the injection pump.

Check 13. Blockage or restriction
in the exhaust system.

A blocked exhaust system is not a common fault and is unlikely, especially if you know that the engine has been running recently.

Corrosion can cause internal parts of silencer boxes to break up and block the system. However, this is only likely with systems that have been in use for a long time. If the system has *not* been used for a long time, mice and birds can make nests in an exhaust pipe. Although the inhabitants may be long gone, the nest material can create blockages.

FAULT 1:
ENGINE WON'T TURN

A seized engine will invariably need a strip down to identify the cause and rectify the problem. Removing the cylinder head is the minimum requirement to allow an initial analysis of the problem. It may be possible to carry out this task in-situ, but removal of the engine is recommended.

IMPORTANT NOTE: ENGINE REBUILDING
Although this sequence appears as '**Job 1**' it relates to several more areas than those shown in **Check 1. Engine Seized**. It is relevant to all faults caused by general engine wear.

Job 1. Engine seized.

The following sequence illustrates a strip down on a modern, small diesel engines. Although the general principles apply to all engines, large or small, you should always consult the manufacturer's workshop manual for details specific to your engine. This strip-down sequence assumes that oil and coolant have been drained from the engine, the battery has been disconnected, the exhaust manifold has been removed and the engine is on a bench or stand.

☐ **Step 1. Component layout.**

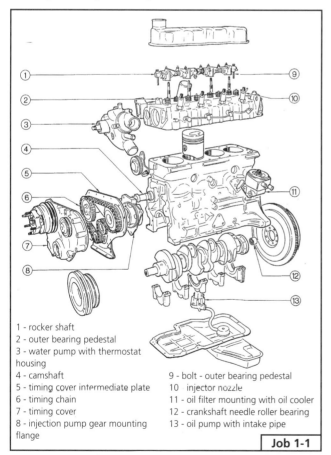

1 - rocker shaft
2 - outer bearing pedestal
3 - water pump with thermostat housing
4 - camshaft
5 - timing cover intermediate plate
6 - timing chain
7 - timing cover
8 - injection pump gear mounting flange
9 - bolt - outer bearing pedestal
10 - injector nozzle
11 - oil filter mounting with oil cooler
12 - crankshaft needle roller bearing
13 - oil pump with intake pipe

Job 1-1

1. Obtain the relevant workshop manual for your engine and study the layout. This drawing shows the names of a typical engine's main components.

• *Even the smallest engine is heavy and one that has been in use for a length of time is likely to be dirty and oily.*

• *It's a good idea to clean a dirty engine before starting work.*

• *Not only is it more pleasant and safe to work on, but there is less chance of getting dirt into places where it can do damage.*

• *Nuts and bolts may be tight and corroded, necessitating high forces to loosen them.*

• *Ideally the engine should be mounted on a stand for strip-down but, at least, it should be securely mounted and supported on a stout workbench - or placed on the floor with a clean board beneath it.*

☐ **Step 2. Remove the cooling fan.**

Job 1-2

2. Watch out for spacers and for bolts of unequal length, which must *always* go back in their correct locations.

☐ **Step 3. Valve rocker components.**

3A. Remove the rocker cover.

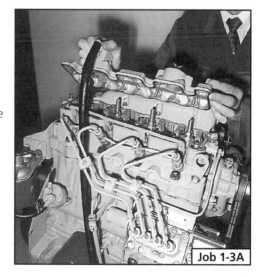

Job 1-3A

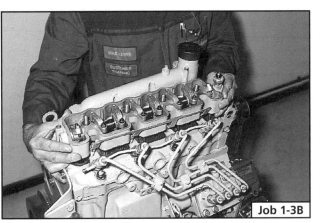

Job 1-3B

3B. Remove the rocker assembly and examine for any damage. On this engine the rocker assembly is built into a housing. If you dismantle the rocker assembly, keep a note of where all the components have come from, so they can be put back in their original locations.

❏ Step 4. Pushrod removal.

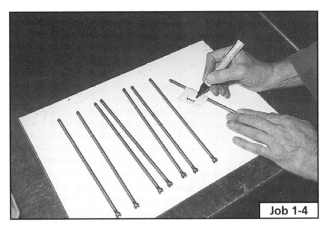

Job 1-4

4. Remove the push rods and look for any that are damaged or bent. To check for straightness, roll the rods along a flat surface. Label the push rods, or keep them in order, so they can be refitted in their original locations.

IMPORTANT NOTE: On many modern engines covers, cover plates, accessories and ancillaries are held in place by studs and bolts that are designed to do their job - and no more! Threads can be stripped or studs and bolts can be sheared very easily, so be very careful if nuts are difficult to loosen, and do not overtighten during reassembly. Apply the manufacturer's recommendations for correct tightening torques.

❏ Step 5. Glow plugs.

5. Remove the electrical connection from the glow plugs (if fitted). Keep a note of how electrical connections are connected together and label the wires if there is likely to be any confusion. This engine uses an uninsulated

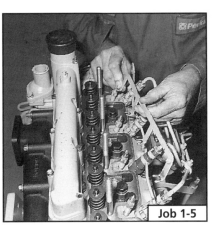

Job 1-5

"bus-bar" so, if your engine has a similar system, make sure there is no danger of this being shorted to earth/ground by ancillary equipment or tools.

❏ Step 6. Fuel pipes.

Job 1-6A

6A. IMPORTANT NOTE: Remove the high pressure fuel pipes from the pump to the injectors. The connectors at either end of the pipes are best loosened with a split-ring spanner (arrowed). Do not bend or kink the fuel pipes. They are subjected to high pressure and can be seriously weakened if distorted.

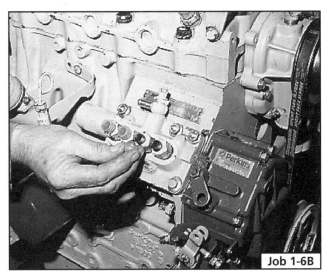

Job 1-6B

6B. Dirt is death to fuel pumps and injectors, so blank the ends of the pipes and pump outlets with caps or tape.

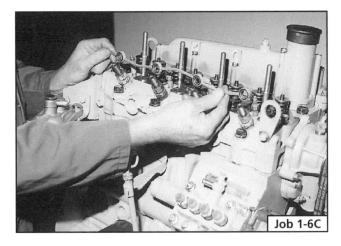

Job 1-6C

6C. Remove the low pressure fuel return pipe from the injectors.

❏ **Step 7. Remove the injectors.**

Job 1-7

7. Use a deep socket, a box spanner or a ring spanner, rather than an open-ended type. Injectors should be kept clean - label them and place them inside a clean plastic bag.

❏ **Step 8. Remove ancillaries.**

8A. Remove the thermostat housing and thermostat. If the engine has been running cold, or overheating, the thermostat can then be checked for correct operation.

Job 1-8A

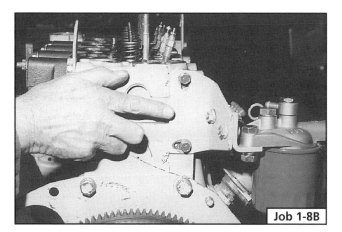

Job 1-8B

8B. Check to make sure that all ancillaries are removed, such as this bracket, connected to both cylinder head and block.

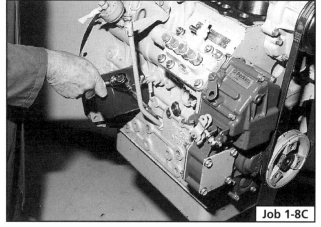

Job 1-8C

8C. Ancillaries such as the dipstick holder can be removed to prevent damage and improve access to other components.

❏ **Step 9. Remove manifolds.**

9. Undo the bolts or nuts retaining the inlet and exhaust manifolds.

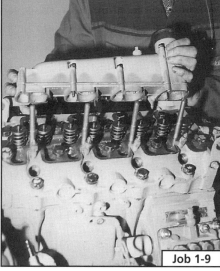

Job 1-9

IMPORTANT NOTE: Make a note of where bolts come from as some can be different lengths and could cause damage if used in the wrong position.

SAFETY FIRST!

• *Cylinder head gaskets often contain asbestos. Read the precautions relating to asbestos in* **Chapter 1 - Safety First!**

❏ **Step 10. Remove cylinder head.**

IMPORTANT NOTE: On some engines, the injector pump will have to be removed first.

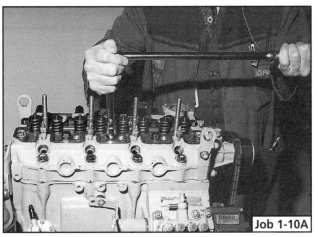

Job 1-10A

10A. Undo the cylinder head nuts or bolts. Consult the manufacturer's manual for the correct sequence. If no information is available it is normal to loosen from the centre out, loosening each bolt a little at a time. Again note where any bolts come from as lengths vary.

10B. Lift off the cylinder head. The head is heavy and will be often be stuck to the block via the gasket. On large engines, it will be necessary to get help or to use a hoist to lift the head off.

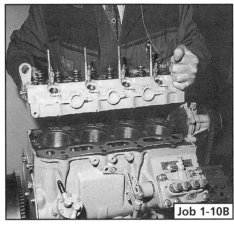

Job 1-10B

making it easy
• *A soft face mallet may be used to loosen the head.*
• *A length of wood, such as the handle of a hammer, may be used in the inlet or exhaust ports to lever the head away from the block.*
• *With injectors in place (fuel feed disconnected) try turning the engine by hand, to see if the compression will break the seal.*

❑ **Step 11. Remove fuel injection pump.**

11A. There are three main types of pump - distributor, in-line and unit injectors. You should consult the manufacturer's manual for specific instructions on pump removal.

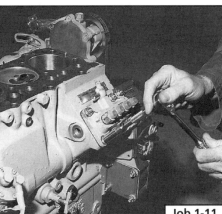

Job 1-11

The type shown here is a variation of the in-line pump known as a cassette pump. (See *Chapter 4* for more information on pump types.)

Job 1-11B

11B. The cut-off solenoid on this engine has to be removed before the fuel injection pump can be taken off.

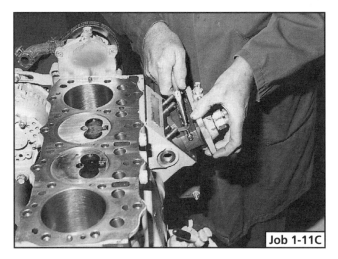

Job 1-11C

11C. 🛈 INSIDE INFORMATION: • Watch out for any internal connections that have to be dismantled before the pump can be completely removed.
• Make a note of where everything goes - it's easy to forget when it comes to reassembly! 🛈

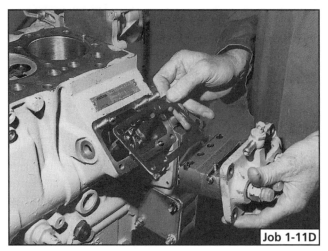

Job 1-11D

11D. Do not throw away the shim if fitted. This one looks like a gasket but different thicknesses are used here to adjust the injection timing. The fuel pump is best placed in a clean plastic bag to prevent it from getting dirty.

❏ **Step 12. More ancillaries.**

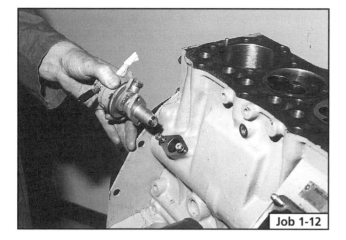

Job 1-12

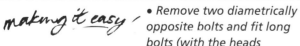

making it easy • **12.** *Now is a good time to remove any ancillaries that are still fitted e.g. the alternator and the fuel lift pump. Here, the bolts for the lift pump have been fitted back into position a few turns. This is good practice, as it prevents them getting lost or confused with others.*

❏ **Step 13. Flywheel removal.**

13A. Undo the bolts retaining the flywheel and remove the flywheel. If the engine is from an automotive application, it may be necessary to remove the clutch cover and friction plate before you can get at the flywheel bolts. The flywheel bolts will be tight so, even if the engine has seized, use an appropriate sized socket or ring spanner on the crankshaft pulley bolt to prevent rotation.

Job 1-13A

making it easy • *Remove two diametrically opposite bolts and fit long bolts (with the heads removed) or studs in their place. These will prevent the heavy flywheel falling as it is loosened. They can also be used later to lock the crankshaft while the pulley bolt is being removed.*

13B. Lever the flywheel off. Take great care, because it may be very tight and suddenly spring loose.

Job 1-13B

❏ **Step 14. Remove the crankshaft pulley.**

Job 1-14A

14A. Lock the crankshaft and loosen the crankshaft bolt. Use a puller to remove the pulley. The type shown is recommended as the leg-type can damage the pulley rim.

14B. Don't forget about the half-moon (Woodruff) key on the crank-shaft (arrowed). It is often overlooked and easily lost.

Job 1-14B

☐ **Step 15. Remove the front cover and timing gear.**

Job 1-15A

15A. Undo the front cover bolts.

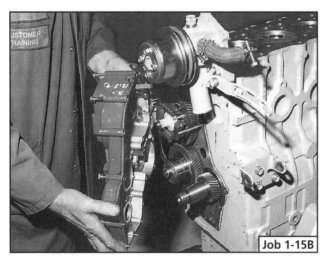

Job 1-15B

15B. Remove the front cover, tapping it with a soft-faced mallet if it 'sticks' to the gasket.

Job 1-15C

FACT FILE: TIMING CHAIN OR GEAR COVERS

15C. Once again, the bolts are usually of different lengths, so keep a note of where they came from. On this engine the camshaft is gear driven. The gears have to be lubricated, so the join between the front cover and cylinder block must be oil tight. There's a gasket between the cover and the block, and a crankshaft oil seal. A similar arrangement is used for chain driven camshafts, but on engines fitted with toothed belt camshaft drives, the front cover is usually just a plastic shield without seals.

making it easy • *Always replace the oil seal, and the gasket. The cost is far less than a strip-down in future.*

Job 1-15D

15D. On this engine, the mechanical governor is fitted to the front of the camshaft. Since designs vary enormously you should consult the manufacturers literature for specific guidance at this stage but, in general, only dismantle as much as you need to remove the camshaft. Here, a plastic slide is all that needs to be removed.

IMPORTANT NOTE: Correct governor operation is essential for satisfactory engine running. Any sticking, stiffness or sloppiness in the mechanism can make it impossible to accurately control engine speed. Consult the manufacturer's manual for assembly instructions.

Job 1-15E

15E. This camshaft is retained by a plate behind the governor drive gear. A hole in the gear allows access to the bolts holding the plate in place, which means turning the gear until the bolt head is visible (arrowed). If the engine is being stripped down because of seizure, it will be necessary to remove the idler gear first. See **Step 17A**.

Job 1-15F

15F. The retaining plate slides out once the screws are removed.

☐ **Step 16. Camshaft removal.**

Job 1-16A

16A. Before the camshaft can be removed, the followers need to be lifted out. Sometimes they are easy to reach with a finger, but with this engine a hook is needed. Again, keep the followers in order so they can be put back in their original locations.

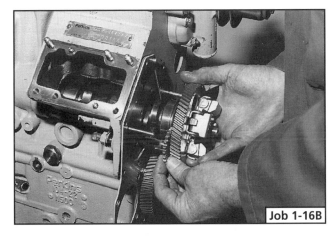

Job 1-16B

16B. The camshaft, and in this case the tachometer drive gear, can now be removed.

IMPORTANT NOTE: Be careful! On engines fitted with plain camshaft bearings, it is very easy for the cams to score the bearings as the camshaft is withdrawn from the engine.

☐ **Step 17. Front plate removal.**

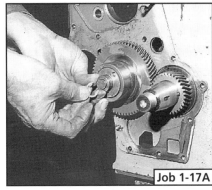

Job 1-17A

17A. A "C" clip is used to lock this idler gear in place.

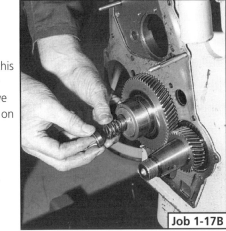

Job 1-17B

17B. Be careful this doesn't fly away when you remove it, especially as - on this Perkins 100 Series engine - there's a spring under the cover.

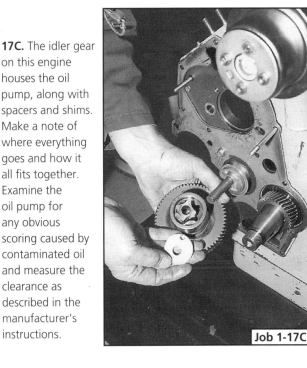

Job 1-17C

17C. The idler gear on this engine houses the oil pump, along with spacers and shims. Make a note of where everything goes and how it all fits together. Examine the oil pump for any obvious scoring caused by contaminated oil and measure the clearance as described in the manufacturer's instructions.

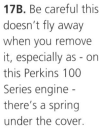

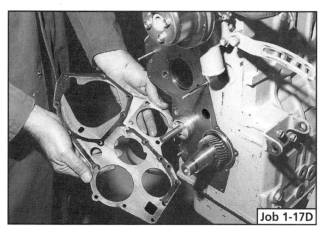

17D. Many engines use adapter plates to suit different applications so shapes can differ greatly. The gasket behind this front plate has been damaged, but, in any case, a new one should be used as a matter of course during re-assembly.

18B. The oil pump strainer and pick-up should not be difficult to remove, but it's wise to make a note of where everything goes. Note the push-fit O-ring seal (arrowed) on this pick-up pipe.

17E. The oil seal behind the rear plate should also be renewed during assembly.

18C. Loosen the big end nuts and remove the big end bearing caps. Sometimes they can be difficult to split, in which case, try rotating the crankshaft gently first one way then the other.

❑ **Step 18. Crank and piston removal.**

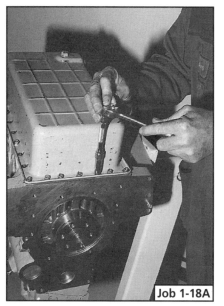

18A. Size of engine permitting, turn the engine over and remove the sump. If a stand is not being used, the top surface of the block can easily be scratched or damaged - use a clean surface.

18D. Remove the bearing caps, looking out for damage or wear at the same time.

Job 1-18E

18E. Keep the caps strictly in order because the caps are matched to the con-rods. If they are not already marked, mark them now, as shown.

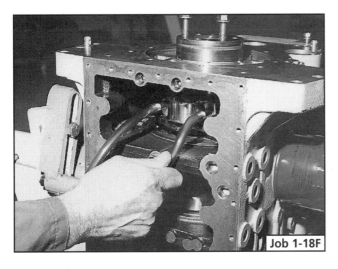

Job 1-18F

making it easy • **18F.** Fitting pieces of plastic or rubber tube over the big end bolts can prevent damage to the crankshaft journals as the pistons and con-rods are removed.

Job 1-18G

18G. Use a rod on the flat mating surface of the con-rod to drift the piston out of the bore. This should not require much force unless a piston has seized. Be careful not to use excess force or cause damage.

IMPORTANT NOTE: Never, unless specifically instructed in the manufacturer's manual, try to remove a piston by driving it towards the crankshaft. It will normally jam against the crankshaft just after the piston rings have escaped from the bore and expanded to prevent the piston going back in again. On a conventional engine with separate main bearing caps this is not too critical - simply remove the caps and the crankshaft. On the engine shown here, which has a "tunnel" assembly it could be impossible to move the piston or remove the crankshaft without smashing the piston rings or causing damage.

Job 1-18H

18H. Remove the pistons and con-rods and keep them in order with their associated big-end caps.

Job 1-18I

18I. Remove the main bearing cap bolts, or the main bearing holder bolts.

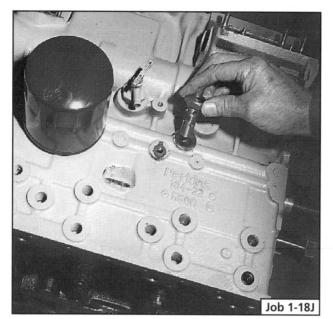

Job 1-18J

18J. Check to make sure that there are no protrusions that would prevent the crankshaft moving. Here the oil pressure relief valve has to be removed before the crank will come out.

Job 1-18K

18K. Some force may be necessary to free the crank. If so, replace the crankshaft pulley nut to prevent thread damage and use a soft faced hammer or, if you haven't got one of these, a standard hammer with a block of wood between the hammer and the nut.

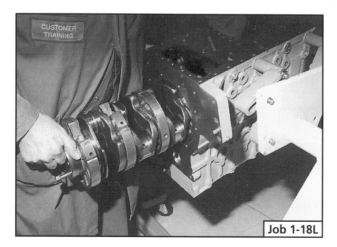

Job 1-18L

18L. On most engines, the crankshaft can be lifted away but on this Perkins 100 Series design, the crankshaft, complete with its bearing holders have to be withdrawn through the block.

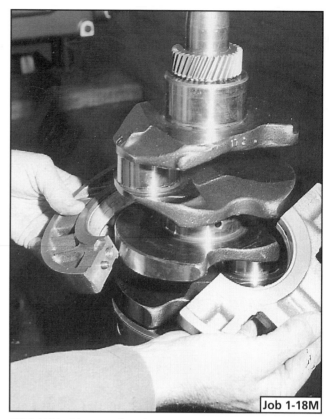

Job 1-18M

18M. With this type of engine, you can only now split the main bearing holders, looking for any damage to journal and bearing shells. Wear, or problems such as journal ovality, can be found by accurate measurement. Check your manufacturer's manual for the appropriate dimensions and measuring techniques.

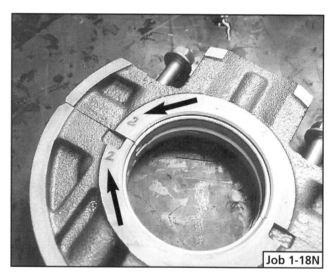

Job 1-18N

18N. Keep the bearing carriers in order and the matching halves together. If the two halves are not already marked (arrowed) you should do so.

❏ **Step 19. Engine inspection.**

Examine the cylinder bores, piston crowns and valves. Catastrophic damage, such as that caused by a stuck valve or a foreign object will be obvious. Scored bores, caused by broken piston rings, dirt or overheating may require closer examination.

In most cases, components will be undamaged. Examine the piston and rings for signs of wear and seizure or for the effects of a hydraulic lock.

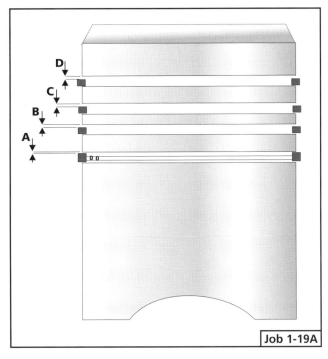

Job 1-19A

19A. Only accurate measurement of piston ring fit and end gap will indicate whether components are serviceable. Piston ring end gaps should be checked with the ring in the bottom (unworn) end of the cylinder. Check your manufacturer's literature for the appropriate dimensions and measuring techniques.

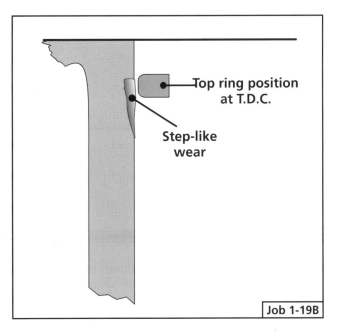

Top ring position at T.D.C.

Step-like wear

Job 1-19B

19B. Bores will need to be measured to check for wear. However, if the wear is really bad a raised lip will be obvious towards the top of the cylinder. Any corrosion may or may not be apparent as this can occur between the piston skirt and cylinder wall.

☐ **Step 20. Engine rebuilding.**

Rebuilding is the reverse of the strip-down process, but precautions are necessary to ensure long and trouble free running. Cleanliness is essential, not just on the visible parts, but also in the oilways, where hidden dirt and grime can be carried to the bearings. Components should be thoroughly cleaned and oilways blown out with compressed air.

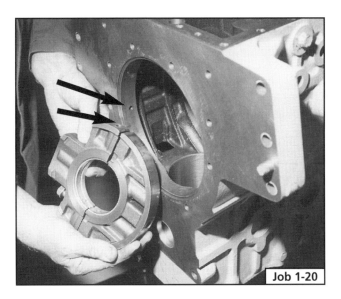

Job 1-20

20. *i* INSIDE INFORMATION: Make sure that oilways match up (arrowed), with gaskets - where fitted - allowing clear passage, otherwise bearings will be starved of oil. *i*

SAFETY FIRST!

• *Compressed air can be dangerous, so never direct it at anybody or at any part of the body. Always wear suitable eye protection.*

Worn and damaged components should generally be replaced although some, such as cylinder bores and crankshafts, can be machined for oversizes. Check the manufacturer's literature for the approved oversizes. Any good engine machine shops should be able to advise on the work that can be carried out.

Job 2. Battery discharged.

☐ **Step 1.** Modern engines have alternators which use sensitive semiconductor voltage control systems so disconnect the battery before charging. Older engines, with dynamo charging systems, are not so critical but if you're not sure it's best to disconnect.

IMPORTANT NOTE: Where the electrical system uses a common earth/ground, such as a vehicle body, it's best to disconnect the earth/ground lead first. If the spanner or screwdriver accidentally touches the body it won't cause a short circuit.

• *The gas given off by a battery is highly explosive. Never smoke, use a naked flame or allow a spark to occur in the battery compartment. Never disconnect the battery (it can cause sparking) with the battery caps removed.*

• *Batteries contain sulphuric acid. If the acid comes into contact with skin or eyes, wash immediately with copious amounts of cold water and seek medical advice.*

• *Do not check the battery levels within half an hour of the battery being charged with a battery charger. The addition of fresh water could then cause the highly acid and corrosive electrolyte to flood out of the battery.*

❑ **Step 2.** A visual check should be made to ensure that there are no cracks, dents, splits or other damage to the plastic casing or electrical terminals of the battery. Damage of this sort may cause either the electrolyte to leak or internal damage, so if serious damage is found, replace the battery.

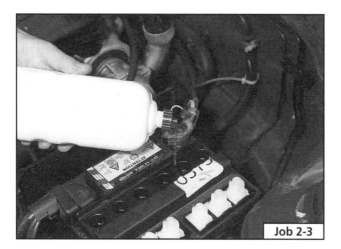

Job 2-3

❑ **Step 3.** Check the electrolyte level in each cell. If the electrolyte level is low, top up with de-ionised or distilled water. Electrolyte should be up to the moulded mark on the battery, or just over the plates, if there is no mark.

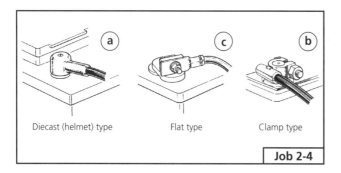

Diecast (helmet) type   Flat type   Clamp type

Job 2-4

❑ **Step 4.** Disconnect the battery. There are two common battery terminal types, (spade and post terminals), but there are three types of connection. A spade connector (**c**) will only fit a spade terminal but a post terminal can either have a clamp (**b**) or a cover (**a**) connection. To disconnect the spade,

completely remove the bolt. A clamp type only needs to have the bolt loosened. If the clamp is difficult to remove, gently ease the clamp jaws apart, don't twist or lever the clamp of as this can damage the battery. The cover type of connection generally has a central screw which has to be completely removed.

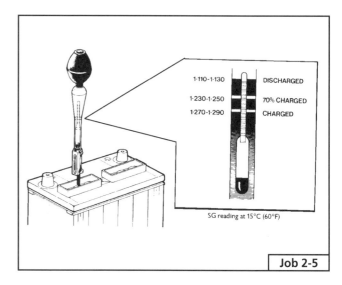

1·110-1·130	DISCHARGED
1·230-1·250	70% CHARGED
1·270-1·290	CHARGED

SG reading at 15°C (60°F)

Job 2-5

❑ **Step 5.** Connect the battery charger making sure that the positive lead goes to the positive terminal and the negative lead to the negative terminal, and switch the charger on. The best way of checking progress is with a hydrometer, which involves drawing up some electrolyte from the cell and checking its Specific Gravity (SG) reading.

Switch the charger off and disconnect it from the battery. Re-connect the battery leads in the reverse order to disconnection.

Job 3. Starter motor jammed.

❑ **Step 1.** Identify the type of electric starter motor fitted, normally either an inertia type (Bendix) or pre-engaged type. On older inertia types a square section of the drive shaft protruded from the end. If you have one of these types, use a suitable spanner to turn the shaft which should free the starter motor. See **Chapter 3, Part B, Check 3**.

❑ **Step 2.** A persistently jamming starter motor indicates a worn flywheel ring gear or starter pinion which can only be permanently repaired by replacing the pinion, starter motor and ring gear. The first step is to disconnect the battery followed by the electrical connections at the starter motor.

❑ **Step 3.** Undo the bolts holding the starter motor to the engine, there are usually three but some motors only have two. Withdraw the motor and inspect the pinion teeth. If these do not appear damaged, have a look at the ring gear on the flywheel.

Step 4. If you are not sure about replacing the pinion yourself, you should consult the appropriate manufacturer's instructions or ask a specialist workshop or garage to carry out the work.

Step 5. If the flywheel ring gear is damaged, the engine will have to be lifted out and the flywheel removed. See *Job 1*. The flywheel can - very occasionally - be rotated to a new position on the crankshaft or a specialist workshop or garage can be asked to replace the ring gear or supply a new flywheel.

Job 4.
Starter circuit wiring faulty.

A multi-meter or other circuit tester, such as a test lamp, aided by a methodical approach should soon identify any wiring faults.

Step 1. Identify the high-and low-current starter circuits. Low-current wires will be thin and route from the battery, via the starter switch, to the starter solenoid. High-current cabling is thick and routes directly from the battery to the starter motor, via the solenoid. See *Chapter 3, Part B, Fig 4*.

Step 2. Disconnect the small diameter wire from the solenoid, connect a multi-meter or test light between end of the disconnected wire and earth/ground. Operate the key switch, push button or whatever actuates the starter motor, and the voltmeter should read the battery's voltage or the lamp should light. If not, then you have a fault in the low-current (switching) circuit. Work back towards the battery, checking the voltage at connectors along the way until the break in the circuit is found. Repair or replace the faulty connector or switch.

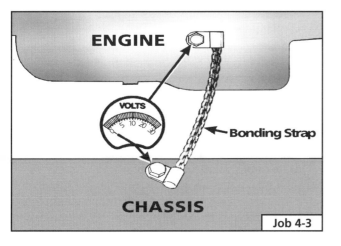

Job 4-3

Step 3. The connections on the heavy-duty cable running between battery and starter motor should be checked for cleanliness and tightness, and the starter motor retaining bolts checked for tightness. Check also the connections on the earth/ground bonding strap (if fitted). The earth/ground bonding strap connects the engine and chassis or baseframe etc. and is designed to ensure that there is an adequate heavy duty earth/ground return between engine and battery. Check the voltage between the relevant terminals, but without disconnecting the heavy-duty cable.

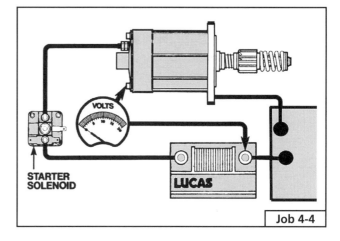

Job 4-4

Step 4. Now carry out a voltage check of the high-current circuit, similar to that in *Step 2*.

Step 5. Repair or replace any faulty wiring or connections found.

Job 5.
Starter motor solenoid faulty.

Step 1. Identify the type of solenoid fitted, either a separate solenoid bolted close to the starter motor, normally with inertia starters, or a solenoid mounted on the starter motor.

Step 2. Disconnect the battery and then the solenoid connections. Remove the solenoid and replace it with a new or serviceable, used solenoid. If the solenoid is mounted on the starter motor, it will be necessary to remove this first. See *Chapter 3, Part C, Job 3*.

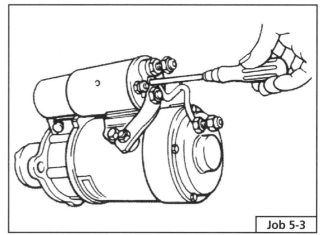

Job 5-3

Step 3. Reconnect the wiring in the reverse order to disconnection. The new solenoid may need to be adjusted - follow the manufacturer's manual, but ONLY adjust the solenoid with the 12v or 24v supply disconnected.

Job 6. Starter motor faulty.

❑ **Step 1.** Disconnect the battery and then the electrical connections to the starter motor.

❑ **Step 2A.** Remove the starter motor and replace with a new motor or serviceable used motor.

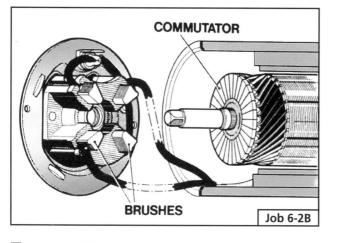

COMMUTATOR

BRUSHES

Job 6-2B

❑ **Step 2B.** *i* INSIDE INFORMATION: The commonest form of starter wear is worn brushes. There are many different types but most require new leads to be soldered in to place. *i*

❑ **Step 3.** Reconnect the wiring in the reverse order to disconnection. Check the drive belt condition and make sure that it is properly tensioned.

FAULT 2: ENGINE TURNS BUT WON'T START

Job 7. Slow cranking speed.

The most common cause of slow cranking speed is a discharged battery. See **Check 2,** as the causes and cures for a partly discharged battery are similar to those for one that is fully discharged. However, if tests show the battery to be in good condition, it is worth checking to see whether there are other reasons why the engine cranks slowly. Check the electrical connections between the battery and the starter motor and make sure they are in good condition.

❑ **Step 1.** Make sure that any load is disconnected from the engine; trying to start a car in gear has an obvious effect, but the application of load may not be so obvious with some marine or industrial engines.

❑ **Step 2.** Check each of the following points:
• Connections from the battery to the starter motor should be cleaned so that bright metal shows. Pay particular attention to battery connections that use a cap over the terminal post - corrosion can build up unseen.
• When reconnecting, make sure that all the connections are tight.

• Investigate the cause of a discharged battery. See **PART B, Check 2**.
• Check for the presence of a decompressor and, if fitted, use it.

IMPORTANT NOTE: Some engines have a decompressor fitted which, by venting the cylinders during cranking, allows a high cranking speed to be built up. When a suitable speed is reached the decompressor is released and the inertia in the flywheel keeps the engine turning at the higher speed long enough for it to start. If a decompressor is fitted, make sure it is used - and in accordance with the maker's instructions.

i INSIDE INFORMATION: If the oil in the sump is the wrong grade the extra load can slow the engine by enough to prevent it from starting - especially in winter time. In a genuine emergency, such as on a sea-going vessel with a problem, while at sea, it may be possible to thin engine oil using diesel or paraffin. An aerosol or spray starting aid can be tried. However this action is drastic and can lead to severe and catastrophic engine damage. If there's no emergency need, don't even try it! Replace the oil as soon a safe to do so. *i*

Job 8. No fuel at injectors.

❑ **Step 1.** Don't rely on the fuel gauge! Always physically check the fuel tank either by visual inspection or by dipping. If the tank level is low it is worthwhile adding some fuel. Even five litres can make a difference as sometimes not all the fuel in the tank is usable. Some installations may have a stop-cock at the fuel tank, so its worth checking to make sure this is open.

❑ **Step 2.** If neither of these two solutions work, some dismantling is necessary as the other most common problem is air in the fuel system and this requires the system to be bled.

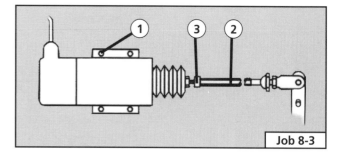

Job 8-3

❑ **Step 3.** Check the operation of the fuel shut-off solenoid (**1**). A faulty fuel shut-off solenoid will prevent fuel from flowing through the pump. Some act directly, in which case the solenoid fits into the injection pump; others operate through linkages. Electrical checks to make sure there are no faults in the electric supply should be carried out. Direct acting solenoids can be felt or listened to, or better still removed to check for correct operation of the plunger.

It should be fairly obvious whether remote types are working but the linkage should be checked to make sure it is set up correctly. To adjust the type shown here, slacken the lock nut (**3**) and rotate the rod (**2**) until correct. Consult the manufacturer's manual for the correct setting procedures.

❑ **Step 4.** Carry out the following checks:
• If the weather is especially cold, and you know that the engine has recently run satisfactorily, there's a good chance the problem is due to fuel waxing - especially if the fuel was put in the tank during a warmer time of the year. (Refineries add anti-waxing compounds in the winter, in cold climates.). Try warming the filter and fuel system with rags soaked in hot water, a hot air blower, hair dryer, or some other safe source of heat.
• In warmer weather, air in the system is the most likely cause. However, check the cut-off solenoid to make sure this is working properly before carrying out further checks.
• It is worth bleeding the fuel system to make sure no air is present even if you're sure the engine had been running satisfactorily and the tank wasn't allowed to run dry. At least you'll be absolutely sure there's no air in the system.
• If bleeding does not improve matters, check for a lift pump fault or a blocked line or filter.
• Having checked the low pressure side, the next check is to see whether anything is coming out of the injectors. A very fine spray from the injector is probably right. Large drops or a dribble probably mean that the pump is not working or there is a blockage on the high pressure side.
• Other faults could be an injector nozzle not sealing correctly, or a restriction or blockage at the injector inlet.

FACT FILE: FUEL CUT-OFF LOCATION
• On rotary pumps, the stop solenoid cuts off the fuel between the transfer pump and the metering valve.
• On in-line pumps, it moves the rack to cut off the fuel supply.

IMPORTANT NOTE: Some diesel fuel injection pumps are self priming and do not have a bleed screw. Check with the engine supplier or pump manufacturer's manual. Randomly undoing screws on a fuel pump can cause catastrophic damage.

❢ INSIDE INFORMATION: To avoid fuel waxing, use a fuel specifically blended for cold weather operations. In a dire emergency, such as at sea with a non-running engine, a solution can be to mix a very small quantity of low grade petrol with the diesel fuel. However, this is a drastic move and should only be used as a last resort. ❢

Looking at the spray from an injector does not tell the whole story about injector condition. Specialist test equipment is needed which can check for:
• Opening Pressure
• Seat tightness
• Back leakage
• Atomisation

EMERGENCY FUEL SUPPLY

If checks reveal a blocked fuel line to the pump, it may be possible, in an emergency, to set up an alternative feed. A possible set up could have the emergency supply feeding direct into the injection pump but it could be set-up to feed anywhere into the low pressure supply. The location will depend on the fault. As long as the emergency supply is kept above the injection pump and topped up with fuel, the engine will continue to run. Remember, this is an emergency supply so the risk of dirt getting into the injection pump when set-up like this must be accepted.

❑ **Step 5.** You may now suspect a faulty injection pump. There are many different types of fuel injection pump. You should identify the pump type and consult the manufacturer's workshop manuals before removing or dismantling a fuel injection pump. It is invariably best to leave this work to a specialist repairer.

❑ **Step 6.** A blocked fuel line or injector will probably have become apparent from the checks already carried out. The important work is to determine the cause of the blockage to prevent reccurrence. It takes quite a lot to block a fuel line so gross fuel contamination or accumulated debris in the tank should be suspected. The fuel filters should prevent any contamination getting to the injectors so, if a blocked injector is found, suspect debris that has got into the system during maintenance work on the delivery side of the filters.

Job 8-7

❑ **Step 7.** A fuel lift pump (if fitted) can wear and become ineffective. The cam lever (arrowed) can wear - but a problem with the diaphragm is the most likely problem. Remove, check and, if necessary, exchange the unit.

❑ **Step 8.** Finally don't overlook blocked filters. They should be changed at the intervals specified by the manufacturer.

Job 9. Water in fuel.

❏ **Step 1.** Check the fuel supply system and locate any water separators or drain points. Examine them and operate the drains.

❏ **Step 2.** If water is present you'll have to drain the fuel system until you are sure that all the water has been removed. Be careful a lot of water in the separators may be due to a build up over a long period of time.

Job 10. Low compression.

❏ **Step 1.** If necessary, remove the cylinder head, see *Chapter 3, Part C, Job 1*. Remove the valves and examine the valve seats. The valve seats and the valve mating surfaces should be even all the way round with no signs of burning or pitting. This is also a good time to check for excessive valve stem wear.

Clean the valve stems and give them a very light smear with oil. Insert them into the guides and check for any side to side movement. Appreciable play indicates either valve stem or valve guide wear. Use a micrometer to check the valve stems for both wear and ovality. You'll need to check the manufacturer's data for acceptable diameters. If the valve stems are within tolerances, any sloppiness will be due to worn guides which will have to be replaced.

❏ **Step 2.** Light pitting or unevenness can sometimes be removed by lapping the valves (especially with older engines) but heavy pitting will require the seats to be recut. Consult the manufacturer's workshop manual for information and limits on recutting.

making it easy ● More modern engines may use hardened valve seats which are scarcely touched by lapping. Slight deterioration of valves or seats can be machine ground out, or if things have gone too far, valves will have to be replaced and seats cut out and replaced with inserts.

❏ **Step 3.** *i* INSIDE INFORMATION: If the valves are satisfactory, have a look at the remains of the head gasket. It will probably have stuck to both the cylinder head and cylinder block so may be in many pieces. The bits you're interested in are the metal sealing rings around the cylinders. Look for signs of gas seepage, usually indicated by dark staining. If there is any present, the head and/or block may need re-facing. *i*

❏ **Step 4.** A blown head gasket should be replaced, making sure that all traces of the old gasket have been removed and the mating surfaces of the cylinder block and head are flat.

❏ **Step 5.** See *Job 1* for further details of engine dismantling and examination.

Job 12. Cold start aids not working.

❏ **Step 1.** If the engine shows no inclination to start, the problem is probably common to all the glow plugs, such as a failure of the relay or time switch. Connect a voltmeter between the glow plug supply and an earth/ground point and activate the start switch. A reading indicates a problem with all the glow plugs. They are not repairable and can only be replaced.

❏ **Step 2.** If there is no voltage at the glow plugs locate the relay and identify the terminals. First check the input from the timer, connect the voltmeter to the appropriate terminal and a good earth/ground point and operate the start switch. A voltage indicates the timer is working but the relay solenoid has failed.

❏ **Step 3.** If there is no current at the relay, the input to the timer should be checked as described in *Step 2*. Before assuming the components are at fault do make sure the wires and cables are properly connected and that none of the connectors or terminals are corroded.

❏ **Step 4.** If the voltage checks indicate that a component is not working, there is little you can do except replace the apparently faulty item.

Job 13. Blockage or restriction in the exhaust system.

A blocked exhaust is an obscure reason for an engine not to start but it can happen and is more likely with an old system or one that has not been used for a long time.

❏ **Step 1.** A visual inspection of the exterior will give some indication of the exhaust system's age. If you aren't familiar with the engine, try to find out when it was last used. Some systems are short and it's easy to check the internal for blockage by prodding with a wire or rod. Some silencers do have intentional restrictions in them so this is not a foolproof check.

❏ **Step 2.** If nothing is found during the *Step 1* checks, dismantling is the only certain way of checking the system - a last resort when all else has failed!

CHAPTER 4
RUNNING PROBLEMS

There are many faults which can prevent an engine from starting - but many more will cause it not to run properly. So once again, don't leap to assumptions - the comments made at the start of Chapter 3 apply just as much to engines that start but do not run as well as they should. Use a methodical and systematic approach which will, in the longer term, save time and money and increase your chances of finding that fault.

Also, remember that the great majority of failures are caused - for one reason or another - by lack of fuel at the

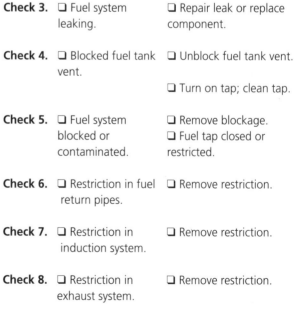

injectors, so do start by checking that there is sufficient usable fuel in the tank. This is particularly important for engines that may be started on a flat or a calm surface but are then used on sloping ground or a rough sea. Fuel sloshing around in the tank may uncover the outlet leading to the pump and let air into the fuel system.

IMPORTANT NOTE: Before replacing any costly components, it is advisable to eliminate all faults caused by less expensive failures or problems.

How To Use This Chapter

This Chapter is divided into three parts:
- PART A lists possible faults and remedies.
- PART B shows you how to check for each fault.
- PART C deals with how to carry out necessary repairs.

Each fault has the same 'Check' number in PART A and PART B, and the same number (now a 'Job' number) in PART C.

Chapter Contents

	Page No.
PART A: FAULT FINDING CHECKLISTS	4-1
PART B: FAULT FINDING STEP-BY-STEP	4-5
PART C: REPAIRS AND MAINTENANCE	4-31

PART A: FAULT FINDING CHECKLISTS

SAFETY FIRST!

• *Industrial and marine engines often have emergency stop devices which will shut the engine down in the event of an emergency. If an engine has one of these devices fitted, you should always investigate the emergency shut-down system before assuming a fault with the engine. Do not attempt to restart one of these engines until the cause of the emergency shut-down has been investigated and resolved.*

FAULT 1: ENGINE STARTS THEN STOPS

CAUSE	REMEDY
Check 1. ❏ Insufficient fuel in tank.	❏ Physically check fuel level.
Check 2. ❏ Air in fuel system.	❏ Bleed fuel system. ❏ Renew fuel and clean fuel system.
Check 3. ❏ Fuel system leaking.	❏ Repair leak or replace component.
Check 4. ❏ Blocked fuel tank vent.	❏ Unblock fuel tank vent. ❏ Turn on tap; clean tap.
Check 5. ❏ Fuel system blocked or contaminated.	❏ Remove blockage. ❏ Fuel tap closed or restricted.
Check 6. ❏ Restriction in fuel return pipes.	❏ Remove restriction.
Check 7. ❏ Restriction in induction system.	❏ Remove restriction.
Check 8. ❏ Restriction in exhaust system.	❏ Remove restriction.
Check 9. ❏ Sticking injectors.	❏ Repair or replace injectors

FAULT 2: ENGINE MISFIRES, RUNS ERRATICALLY OR SURGES

CAUSE	REMEDY
Check 10. ❑ Restriction in induction system.	❑ Remove restriction.
Check 11. ❑ Wrong or incorrectly connected high pressure pipes.	❑ Renew or reconnect correctly.
Check 12. ❑ Faulty lift pump.	❑ Repair or replace lift pump.
Check 13. ❑ Air in fuel system.	❑ Bleed fuel system.
Check 14. ❑ Restricted fuel feed pipe.	❑ Remove restriction.
Check 15. ❑ Blocked fuel tank vent.	❑ Unblock fuel tank vent.
Check 16. ❑ Incorrect valve clearances.	❑ Adjust valve clearances.
Check 17. ❑ Broken or weak valve springs.	❑ Renew valve springs.
Check 18. ❑ Sticking valves.	❑ Overhaul cylinder head.
Check 19. ❑ Incorrect injection pump timing.	❑ Correct injection pump timing.
Check 20. ❑ Incorrect valve timing.	❑ Correct the valve timing
Check 21. ❑ Incorrect or faulty injectors.	❑ Renew injectors.
Check 22. ❑ Engine does not run at correct operating temperature.	❑ See *FAULT 8: ENGINE OVERHEATS* or *FAULT 7: ENGINE RUNS COLD*.
Check 23. ❑ Poor cylinder compressions.	❑ Overhaul engine.
Check 24. ❑ Speed control linkages stiff or sticking.	❑ Repair or replace linkages.
Check 25. ❑ Faulty stop control operation.	❑ Repair or replace stop control mechanism.
Check 26. ❑ Choked fuel filter.	❑ Renew fuel filter.
Check 27. ❑ Faulty governor.	❑ Overhaul or replace governor.

FAULT 3: ENGINE VIBRATES EXCESSIVELY

CAUSE	REMEDY
Check 28. ❑ Sticking or restricted speed control movement.	❑ Free off mechanism, repair or replace linkages.
Check 29. ❑ Faulty engine mountings.	❑ Repair or replace mountings.
Check 30. ❑ Damaged or loose fan or crankshaft pulley.	❑ Renew or secure fan or pulley.
Check 31. ❑ Wrong or incorrectly connected high pressure pipes.	❑ Renew or reconnect correctly
Check 32. ❑ Sticking valves.	❑ Overhaul cylinder head.
Check 33. ❑ Incorrectly aligned or loose flywheel.	❑ Align or secure flywheel.
Check 34. ❑ Poor cylinder compressions.	❑ Overhaul engine.
Check 35. ❑ Incorrect or faulty injectors.	❑ Renew injectors.
Check 36. ❑ Incorrect injection pump timing.	❑ Adjust injection pump timing.
Check 37. ❑ Incorrect idle speed setting.	❑ Adjust idle speed setting.
Check 38. ❑ Leaking injector pipes (fuel escaping).	❑ Tighten unions or replace pipes.
Check 39. ❑ Incorrect valve clearances.	❑ Adjust valve clearances.

FAULT 4: ENGINE KNOCKS EXCESSIVELY

CAUSE	REMEDY
Check 40. ❑ Incorrect type or grade of fuel.	❑ Drain system and refill with correct fuel.
Check 41. ❑ Incorrect or faulty injectors.	❑ Renew injectors.
Check 42. ❑ Incorrect injection pump or valve timing.	❑ Replace injection pump or adjust timing.
Check 43. ❑ Incorrect valve clearances.	❑ Adjust valve clearances.
Check 44. ❑ Broken valve springs.	❑ Renew valve springs.

Check 45. ❏ Sticking valves.   ❏ Overhaul cylinder head.

Check 46. ❏ Incorrect piston height.   ❏ Overhaul engine.

Check 47. ❏ Worn or damaged small-end or big-end bearings.   ❏ Renew bearings.

Check 48. ❏ Excessive camshaft bearing wear.   ❏ Repair as necessary.

Check 49. ❏ Broken, worn or sticking piston rings.   ❏ Renew piston rings.

Check 50. ❏ Worn or damaged pistons.   ❏ Renew pistons

Check 51. ❏ Excessive timing gear backlash.   ❏ Overhaul or replace timing gear.

Check 52. ❏ Overfuelling through faulty or sticking governor speed control system.   ❏ Overhaul or replace governor or speed control system.

Check 53. ❏ Faulty lift pump.   ❏ Repair or replace lift pump.

Check 54. ❏ Faulty cold starting equipment.   ❏ Repair or replace cold start equipment.

Check 55. ❏ Overheating.   ❏ See *FAULT 8: ENGINE OVERHEATS*.

Check 56. ❏ Piston seizure/ pickup.   ❏ Overhaul engine.

FAULT 5: ENGINE WILL NOT ACCELERATE OR PRODUCE POWER

CAUSE	REMEDY
Check 57. ❏ Stop control not in fully ON position.	❏ Put stop control in fully ON position.
Check 58. ❏ Engine speed control sticking or movement restricted.	❏ Free off mechanism, repair or replace linkages.
Check 59. ❏ Restriction in induction system.	❏ Remove restriction.
Check 60. ❏ Air in fuel system.	❏ Bleed fuel system.
Check 61. ❏ Partly blocked fuel feed pipe.	❏ Remove blockage.
Check 62. ❏ Choked fuel filters.	❏ Renew fuel filters.
Check 63. ❏ Blocked fuel tank vent.	❏ Unblock fuel tank vent.

Check 64. ❏ Faulty lift pump.   ❏ Repair or replace lift pump.

Check 65. ❏ Incorrect injection pump or valve timing.   ❏ Replace injection pump or adjust timing.

Check 66. ❏ Restriction in exhaust system.   ❏ Remove restriction.

Check 67. ❏ Faulty, worn or incorrect injectors.   ❏ Renew injectors.

Check 68. ❏ Faulty, worn or incorrect injection pump.   ❏ Overhaul or replace injection pump.

Check 69. ❏ Incorrect valve clearances.   ❏ Adjust valve clearances.

Check 70. ❏ Poor cylinder compressions.   ❏ Overhaul engine.

Check 71. ❏ Poor boost pressure (turbo charged engines).   ❏ Overhaul boost control system.

Check 72. ❏ Sticking fuel delivery valves.   ❏ Overhaul or replace injection pump.

FAULT 6: EXCESSIVE LUBRICATING OIL CONSUMPTION

CAUSE	REMEDY
Check 73. ❏ Oil leak.	❏ Repair leak.
Check 74. ❏ Oil level too high.	❏ Drain oil to correct level.
Check 75. ❏ Engine breathing system blocked.	❏ Unblock breathing system.
Check 76. ❏ Worn valve stems or valve guide bores.	❏ Overhaul cylinder head.
Check 77. ❏ Worn valve stem seals.	❏ Renew seals
Check 78. ❏ Worn cylinder bores or pistons.	❏ Overhaul engine.
Check 79. ❏ Worn or broken piston rings.	❏ Overhaul engine.
Check 80. ❏ Wrong type of oil, diluted oil or inferior quality oil.	❏ Drain oil and replace with correct grade.
Check 81. ❏ New or rebuilt engine not fully bedded in.	❏ Run-in engine for period recommended by manufacturer.
Check 82. ❏ Glazed cylinder bores.	❏ Hone bores.

Check 83. ❏ Faulty oil cooler. ❏ Repair or replace oil coolers.

Check 84. ❏ Cross leakage between oil feed pipe and fuel pipe. ❏ Repair or replace leaking pipes.

Check 85. ❏ Oil leaks from ancillary equipment. ❏ Renew oil seals.

Check 86. ❏ Consumption by the fuel injection pump. ❏ Repair or replace injection pump.

FAULT 7:
ENGINE RUNS COLD

CAUSE	REMEDY
Check 87. ❏ No thermostat, faulty thermostat or wrong thermostat fitted.	❏ Fit serviceable thermostat.
Check 88. ❏ Faulty gauge or temperature transmitter.	❏ Repair or replace faulty components.

FAULT 8:
ENGINE OVERHEATS

CAUSE	REMEDY
Check 89. ❏ Insufficient coolant.	❏ Add coolant to the correct level.
Check 90. ❏ Oil level too high.	❏ Drain oil to correct level.
Check 91. ❏ Faulty thermostat.	❏ Fit serviceable thermostat.
Check 92. ❏ Blocked coolant system, faulty or incorrect radiator, coolant hoses or pressure cap.	❏ Flush or clean system, repair or replace faulty or incorrect components.
Check 93. ❏ Loose fan belt.	❏ Tighten fan belt.
Check 94. ❏ Faulty water pump.	❏ Repair or replace water pump.
Check 95. ❏ Leaking cylinder head gasket or cracked cylinder head.	❏ Replace cylinder head gasket or repair and replace cylinder head.
Check 96. ❏ Incorrect injection pump or valve timing.	❏ Replace injection pump or adjust valve timing.
Check 97. ❏ Faulty or incorrect injectors.	❏ Renew injectors.

Check 98. ❏ Restriction in induction system. ❏ Remove restriction.

Check 99. ❏ Restriction in exhaust system. ❏ Remove restriction.

Check 100. ❏ Blocked gearbox or engine oil cooler. ❏ Flush, clean or replace oil cooler.

Check 101. ❏ Faulty gauge or temperature transmitter. ❏ Repair or replace faulty components.

FAULT 9:
OIL PRESSURE TOO LOW

CAUSE	REMEDY
Check 102. ❏ Engine lubricating oil too thin.	❏ Drain and refill with correct oil grade.
Check 103. ❏ Worn or damaged bearings.	❏ Overhaul engine.
Check 104. ❏ Insufficient oil in sump.	❏ Top up to correct level
Check 105. ❏ Faulty gauge or pressure transmitter.	❏ Repair or replace faulty components.
Check 106. ❏ Worn oil pump.	❏ Overhaul or replace oil pump.
Check 107. ❏ Pressure relief valve stuck open.	❏ Overhaul or replace relief valve.
Check 108. ❏ Broken relief valve spring.	❏ Fit new relief valve spring.
Check 109. ❏ Faulty suction pipe.	❏ Repair or replace suction pipe.
Check 110. ❏ Blocked oil filter.	❏ Renew oil filter
Check 111. ❏ Gearbox or engine oil cooler choked.	❏ Flush, clean or replace oil cooler.
Check 112. ❏ Blocked sump strainer.	❏ Clean sump strainer.

FAULT 10:
OIL PRESSURE TOO HIGH

CAUSE	REMEDY
Check 113. ❏ Engine lubricating oil too thick.	❏ Drain and refill with correct grade.
Check 114. ❏ Faulty gauge or pressure transmitter.	❏ Repair or replace faulty components.

Check 115. ❏ Pressure relief valve sticking closed.
❏ Overhaul or replace relief valve.

FAULT 11: ENGINE OVERSPEEDS

CAUSE	REMEDY
Check 116. ❏ Incorrect maximum speed setting.	❏ Adjust maximum speed setting.
Check 117. ❏ Faulty or incorrectly set governor.	❏ Renew or adjust governor.
Check 118. ❏ Faulty or incorrectly set injection pump.	❏ Overhaul or replace injection pump.

FAULT 12: ENGINE WILL NOT STOP

CAUSE	REMEDY
Check 119. ❏ Faulty, damaged or incorrectly set stop mechanism.	❏ Repair, renew or adjust mechanism.
Check 120. ❏ Fuel or oil leak into induction system or cylinders.	❏ Stop oil leaking into induction system or cylinders.

FAULT 13: EXCESSIVE FUEL CONSUMPTION

CAUSE	REMEDY
Check 121. ❏ Fuel leaks.	❏ Repair fuel leak or replace leaking components.
Check 122. ❏ Restriction in induction system.	❏ Remove restriction.
Check 123. ❏ Sticking valves.	❏ Overhaul cylinder head.
Check 124. ❏ Incorrect valve clearances.	❏ Adjust valve clearances.
Check 125. ❏ Faulty cold starting aid.	❏ Repair or replace cold start
Check 126. ❏ Incorrect injection pump or valve timing.	❏ Renew injection pump or adjust valve timing.
Check 127. ❏ Faulty or incorrect injectors.	❏ Renew injectors.
Check 128. ❏ Faulty or incorrect injection pump.	❏ Overhaul or replace injection pump.
Check 129. ❏ Poor cylinder compressions.	❏ Overhaul engine.

PART B: FAULT FINDING STEP-BY-STEP

IMPORTANT NOTE: Where the check is the same as the repair e.g. finding and clearing a blocked pipe, PART B tells the whole 'story'. However, where the discovery of a fault leads to the need for a mechanical repair, see PART C for an overview, and the engine manufacturer's manual for specific details.

FACT FILE: TURNING THE ENGINE BY HAND

• There are often times when it is necessary to turn an engine by hand e.g. making sure that pistons are at the top or bottom of their strokes. Because of the high compressions used, this can be very difficult.

• A suitably sized spanner or socket on the cranknut is the best way to turn the engine and the job can be made much easier by removing the glow plugs, the injectors or, if fitted, operating the decompressor.

• Alternatively, if the engine is fitted in a vehicle, it is often possible to put the engine in gear and turn the engine by pushing the vehicle.

• Sometimes, especially with small engines, the drive belts can provide sufficient friction and these can be used to turn the engine.

SAFETY FIRST!

• *If you have not removed the injectors or glow plugs, do make sure the engine STOP mechanism is activated.*
• *There is always a chance that the engine may fire and run, with disastrous consequences.*

FAULT 1: ENGINE STARTS THEN STOPS

Check 1. Insufficient fuel in tank.

As always, don't rely on the fuel gauge. Some faults in an electric fuel gauge circuit can give very misleading readings and, if precautions are not taken, so can sight gauges and float-based mechanical gauges.

making it easy • A fault in an electric fuel gauge's circuitry can give very misleading indications.

• Most gauges show full when the electrical resistance of the transmitter in the tank is low and empty when the electrical resistance is high.

• If the electrical connection to the tank is broken (open circuited), the tank will show empty while, if there is a short circuit in the supply, the gauge will show full, irrespective of the amount of fuel in the tank.

1A. Deliberately open-circuiting the supply connection at the tank should see the gauge read empty. Shorting it to earth/ground will cause the gauge to indicate full. These checks can give a good indication of the gauge's serviceability, but do check the wiring diagram to make sure you know which is the supply and which is the earth/ground connection.

Always physically check the fuel tank either by visual inspection or by dipping. If the tank level is low it is worthwhile adding some fuel, especially with engines that may not run on level surfaces. On marine engines, a rough sea will cause fuel to slosh around in the tank, possibly uncovering the fuel outlet pipe, while, on road vehicles, negotiating a steep hillside or running an engine on a gradient can have a similar effect, although there may appear to be sufficient fuel in the tank when it is checked on the level. Often, fuel tanks used in these conditions will have more that one outlet, to ensure continuity of supply.

1B. A visual inspection of the fuel level is easier on some tanks than others; and easiest if a sight gauge is fitted. If the filler is directly on the tank it should be easy to see how much fuel is in the tank.

If the route from the filler to the tank is reasonably straight dipping is easy, otherwise a flexible dipstick may be necessary. A wooden dipstick should give a much more visible indication of fuel than one made from metal. Be careful that the dipstick does not drop into the tank!

1C. On tanks with sight gauges - often a glass or plastic tube running vertically up the side of the tank - make sure that any isolating valves at the top and bottom are open. If these are shut during the check all you will see will be the fuel trapped in the tube. With this type of gauge it is very easy to mistake a completely empty tank for one that is completely full, or vice versa. If the tank is completely full, the actual fuel level may be above the top of the sight gauge, which will be completely full of fuel with no fuel level visible. Sight gauge tubing rapidly becomes discoloured so it is often difficult and sometimes impossible to 'read'.

It is good practice to top-up the fuel tank when an engine is shut-down at the end of the day or when it will not be used for a prolonged period. A full tank reduces the risk of water condensation in the tank and ensures you always have a full tank to cover any unplanned engine use.

Check 2. Air in fuel system.

If you have allowed the fuel level to get too low or the fuel tank to run dry, it's fairly certain that air will have entered the fuel system. Some injection pumps are self-priming and should allow continued operation when the fuel tank is refilled. You should check with your manufacturer to see whether you have one of these pumps. However, other parts of the fuel system may not be so tolerant. If you have not allowed the fuel level to get low or the tank to run dry, any air in the system will be due to leaks in the fuel system - check the pipes and unions between the tank and the lift pump inlet.

IMPORTANT NOTE: Do check to see whether you have a self-priming pump. It would be disastrous to undo bolts or screws randomly in a vain search for a bleed screw. Similarly, if you do not have a self priming pump, make sure you know exactly where the bleed points are: Consult the manufacturer's manual.

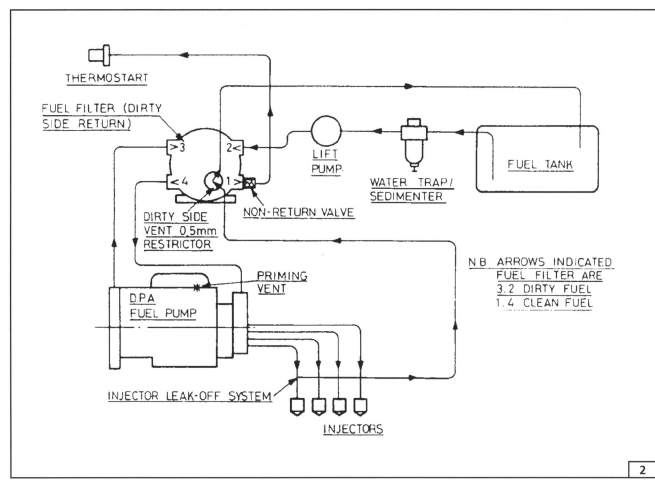

THERMOSTART

FUEL FILTER (DIRTY SIDE RETURN)

>3

2<

<4

1>

LIFT PUMP

WATER TRAP / SEDIMENTER

FUEL TANK

DIRTY SIDE VENT 0.5mm RESTRICTOR

NON-RETURN VALVE

N.B. ARROWS INDICATED FUEL FILTER ARE
3.2 DIRTY FUEL
1.4 CLEAN FUEL

PRIMING VENT

D.P.A. FUEL PUMP

INJECTOR LEAK-OFF SYSTEM

INJECTORS

2

making it easy

2. • Be systematic and work from the fuel tank to the injectors.

• Check each of the unions and make sure that fuel flows freely from pipes.

• When you have reached and, if necessary, bled the injection pump, the high pressure fuel should now force any remaining air from the high pressure side of the system.

• IMPORTANT NOTE: If the connections are secure between the tank and the lift pump, it is unwise to disturb them because this will probably cause an air leak.

• It is better to start removing air from the fuel system at the fuel filter and then move on to the injection pump and then the high-pressure connections at the injectors.

IMPORTANT NOTE: Don't allow the fuel level to get too low and remember that a fuel level that appears sufficient for operation on the level may not be enough in a rough sea or in hilly or rough terrain. Inspect the fuel system at regular intervals to make sure that pipes are not damaged and unions are in good condition.

Check 3. Fuel system leaking.

SAFETY FIRST!

• Although diesel fuel is not flammable at normal temperatures it can form a very flammable combination if absorbed into paper, cardboard, fabrics or any other material that acts as a wick.

• A leak from the high pressure side of the fuel delivery system can produce a flammable spray and the spray can penetrate any exposed parts of the body with potentially lethal consequences.

• On the latest electronic injection pumps, and on common rail systems, where ultra-high pressures are used, do not release a high pressure connection until at least 30 minutes after the engine has stopped. See manufacturer's manual for details.

• If skin should be injected, see medical help immediately.

3A

3A. ℹ️ INSIDE INFORMATION: Leaking components can sometimes be repaired and you should consult the manufacturer or supplier to see if any overhaul kits are available. Damaged pipes and components generally have to be renewed. Ensure that any flexible or plastic pipes are suitable for use with fuel. Tightening a leaking union can often cure the leak but renewal may be the only satisfactory solution. ℹ️

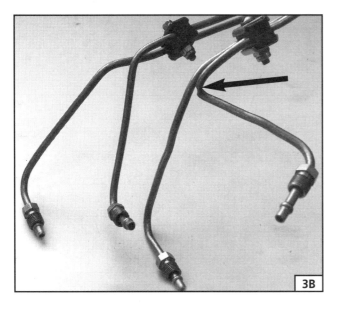

3B

making it easy **3B.** • Although it is frequently possible to repair unions and components, it is often quicker, less expensive and more reliable to fit new or serviceable used components. A perfect example is this section of kinked fuel delivery pipe (arrowed).

• NEVER attempt to repair a high pressure pipe, from pump to injectors. ALWAYS fit a new one.

3C

3C. Leaks are much easier to spot and trace when the engine and fuel delivery system are clean. Wipe down the engine on a routine basis; the less dirt there is around, the smaller the chance of it getting into places where it can do damage.

Check 4. Blocked fuel tank vent.

Some fuel systems have a small vent hole in the filler cap, while others may have pipes or tubes leading from the top of the tank or filler pipes. Some filler caps have vents built in to the edges of the cap.

4. Use some small diameter wire - or compressed air - to unblock vent holes, preferably working from the inside-out if the component cannot be removed, so as not to contaminate the fuel or, if using compressed air, pressurise the system. If the tank has vent pipes, these should be disconnected and blown through to clear any blockage. At the same time make sure the pipe connections on the tank are not blocked.

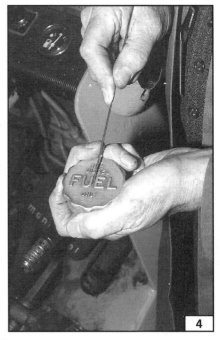

4

ENVIRONMENT FIRST!

• A leaking fuel system wastes fuel and can be environmentally unacceptable, especially on inland waterways where the discharge of contaminated bilge water is prohibited.

• Fuel leaks can also be dangerous with a risk of fire or injection of fuel into the bloodstream.

SAFETY FIRST!

• If you blow through pipes with compressed air, wear suitable eye protection.

• Also, make sure the open ends of any pipes are pointing away from you or anybody else in the area.

Investigate the cause of a blocked fuel tank vent so that you can decided whether this problem is likely to happen again. Unless the environment is extremely dirty, blocked fuel tank vents are uncommon.

Check 5. Fuel system blocked or contaminated.

CHECK FUEL COCK

• First of all, make sure that any fuel cocks are fully open.
• It is not unknown for them to be mistaken for other valves and closed in error especially if the fuel tank is remote from the engine.
• A partially open fuel cock may allow sufficient fuel flow to get the engine started but as fuel demand increases, it will prevent sufficient flow.

SAFETY FIRST!

• *If you blow through pipes with compressed air, wear suitable eye protection.*
• *Make sure the open ends of any pipes are pointing away from you and anyone else in the area.*

CHECK FOR BLOCKAGES OR WATER IN FUEL

• A partial or complete blockage of the fuel feed pipe may be due to an accumulation of debris over a long period of time, or it can indicate gross contamination of the fuel.
• In some situations, debris can be stirred up by allowing the tank to nearly run dry, or by driving a vehicle over rough ground or a boat encountering a heavy sea.
• Similarly with any water that may be in the tank. Any movement can cause water to migrate to the fuel pick-up point.
• If you suspect a blocked fuel line or water in the tank, check to make sure that the fuel is not contaminated otherwise the problem can recur.

making it easy / • *Be systematic and work from one end of the fuel delivery system to the other.*

• *Loosen each union in turn and check to see whether there is a good flow of fuel from the open pipe before retightening. Using this method, you should be able to find the blocked section of pipe.*

BLOCKED FUEL FILTER

5A. If a fuel filter is either partly choked or fully blocked, it will have to be renewed. Find out where the fuel filter or filters are in the system. There should be at least one and possibly two on the inlet side of the injection pump. However, look for any others that may have been put into the system, such as an in-line filter between the tank and lift pump. Consult the manufacturer's manual for the correct replacement type and any special fitting instructions.

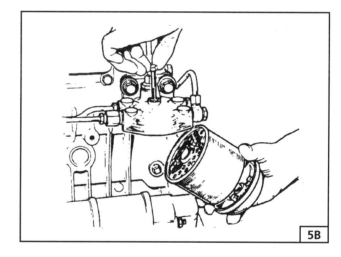

5B. Follow the manufacturer's instructions for renewal and remember that it will be necessary to bleed the system. If the filters are relatively new, you should investigate the fuel system for contamination.

Check 6. Restriction in fuel return pipes.

Check your engine to see what type of fuel return pipes are fitted and where they are routed. Most engines have a common return with all the injectors connected together and a common pipe feeding back into the low pressure system. Sometimes this is via the injection pump but may be direct to the fuel tank or into the fuel filter or lift pump.

The fuel return pipes operate at low pressure and allow excess fuel from the injectors to return to the fuel system. They are sometimes made of plastic or rubber which can be compressed or kinked. Examine the fuel return pipes for any obvious signs of compression or kinks.

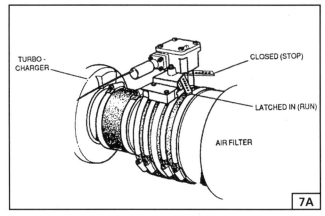

6. Examine the fuel return pipes for any obvious signs of compression, deterioration or kinks. Disconnect the common return pipe at the injector end and try running the engine. Fuel should flow out of the injector connection and, if the engine continues to run, the blockage will be somewhere in the common pipe or the connection at the other end.

• As a further check, reconnect the pipe and disconnect it at the other end. If fuel flows the blockage is in the connection; no fuel, and the pipe is blocked.

• The fuel return pipes are sometimes just push-fits on their connections but most are more securely fixed. (If so, do not confuse them with the high pressure delivery pipes.) Remove each return pipe and use a wire to make sure there are no blockages, (not common!) or blow through them with compressed air.

• Flush out the pipe with diesel fuel before refitting, to make sure that any residue from the blockage is completely removed. Also, check the connectors at each end to make sure these are not blocked.

7A. Check to see if there is an emergency air shut-off valve - see *Chapter 6, Ancillaries* for further information. Do not attempt to start an engine if the valve has been activated; find out why and correct the problem first. If you have any doubts, renew or clean the air filter but do not run the engine without one - it's surprising how much premature cylinder bore and piston ring wear will take place.

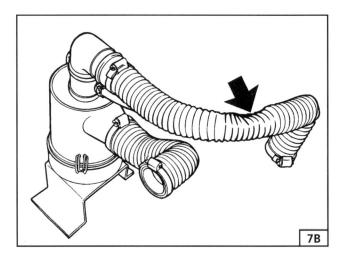

7B. If a new or cleaned filter is fitted and the problem persists, check the intake ducting for both damage (arrowed) and blockages. Convoluted intake pipes can collapse internally - and it's surprising where mice and birds will build their nests!

Change the air filter at the manufacturer's recommended interval, or more frequently if the operating conditions are very dusty, such as around construction work. If a paper element filter is found to be wet, investigate the reason and take action to prevent it happening again.

In an emergency, run the engine without a filter but be aware of the damage that contaminated intake air can cause.

Check 7.
Restriction in induction system.

A blocked air filter is usually due to a build up of dirt or, if you have a paper element type of filter, the filter element becoming wet. Paper element types have to be renewed but other types, such as oil bath or wire gauze filters can be cleaned. Some industrial engines have an emergency air flow shut-off valve. Check to see if there is one fitted and, if it has been activated, find out why before attempting to start the engine. See *PART C* for air filter types.

Check 8.
Restriction in exhaust system.

A blocked exhaust system is not a common fault and is unlikely, especially if you know the engine has been running recently.

8. Corrosion can cause internal parts of silencer boxes to break up and block the system. This is only likely with systems that have not been in use for a long time. If the system has been out of use, mice and birds can make nests in exhaust pipes. Although the inhabitants may be long gone, the nest material can create blockages. Dismantle and check if no exhaust gases pulse out of the pipe as the engine is cranked.

Exhaust systems are often forgotten about until they go wrong, so it is a good idea to routinely inspect the exterior for deterioration. If the outside looks past it, early replacement will prevent problems occurring, which are normally at the most inconvenient time.

Check 9. Sticking injectors.

Unless they are very old and worn or dirt has been allowed to get into them, sticking injectors are unlikely. On a multi-cylinder engine, it is highly unlikely that all the injectors would stick at the same time so misfiring, rather than a complete stop, would be the result. Apart from checking to see whether the injectors produce a fine spray pattern there is little else that can be done, since faulty injectors will have to be renewed or properly serviced.

First, check to make sure that fuel is being delivered to the injectors. Loosen the union between the high pressure pipe and the injector and crank the engine. A good spurt of fuel should be seen. Remove the injector, reconnect the union and crank the engine. A fine spray should emit from the injector.

Assuming fuel is being delivered to the injector, anything other than a fine spray, such as large droplets, dribbles or nothing at all means that the injector should be replaced by a new or serviceable used item. It may be possible to repair the old injector but specialist test equipment is needed to find out what caused the problem and whether the injector is repairable.

To gain the longest possible life from injectors, make sure that the fuel entering the injection pump is properly filtered. If the high pressure pipes are disconnected at any time, ensure that no dirt can get into the system.

FAULT 2: ENGINE MISFIRES, RUNS ERRATICALLY OR SURGES

Check 10.
Restriction in induction system.

See **Check 7**.

Check 11. Wrong or incorrectly connected high pressure pipes.

If the high pressure pipes have been removed and refitted, it is sometimes possible to connect the injector feed pipes to the outlet ports on the injection pump in the wrong positions. The injection will then be out of sequence.

IMPORTANT NOTE: This is not a problem you can encounter with unit injectors as the high pressure is generated at the injector itself.

11. Check the manufacturer's manual for the correct firing order and any guidance on the correct routing of the high pressure pipes. If any pictures or drawings are provided, make sure they refer to the specific engine and pump combination you have. This, for instance, is the drawing referring to the Perkins 12SE Series engine.

If no information is available, it will be necessary to rotate the engine crankshaft in the direction of rotation to find out when each piston in turn is on the compression stroke. See **PART C, Job 11** for further information. As the piston approaches the top of the compression stroke, note which injection pump outlet port is delivering fuel and connect the high pressure pipe from the port to the appropriate injector.

making it easy
- *When you are carrying out any dismantling work, always make a note of where components come from and how they were connected together.*
- *The dismantling takes a little longer but more time is usually saved during reassembly and there is less chance of damaging the engine through incorrect refitting of components.*

Check that the pipes are the correct specification, such as having the correct internal diameter. If not, replace with the correct pipes, connected in the right sequence.

Check 12. Faulty lift pump.

A lift pump may fail completely or only deliver a fraction of the intended quantity of fuel. In the first case, the engine may not run at all while in the second case, misfiring will occur under heavy loads. Some lift pump faults, such as a punctured diaphragm, can often be repaired but you will need to check with the manufacturer or supplier to see if yours is a repairable type and whether repair kits are available.

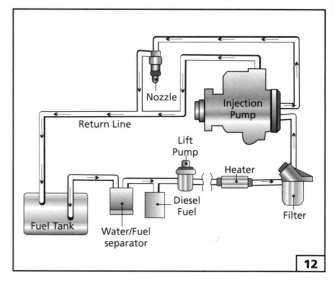

12. Remove the outlet pipe from the lift pump and crank the engine. The pump should deliver good strong spurts of fuel. Unless the manufacturer specifies a delivery rate or pressure at cranking speed, this is the only check you can really carry out without the use of pressure test equipment.

It is sometimes possible to adjust the pressure output of pushrod-operated pumps by changing the number or thickness of spacers. See your operator's manual.

Check 13. Air in fuel system.

See **Check 2**.

Check 14. Restricted fuel feed pipe.

A partial blockage of the fuel feed pipe may be due to an accumulation of debris over a long period of time, or it can indicate gross contamination of the fuel. If you find a blocked fuel line check to make sure the fuel is not contaminated.

making it easy
- *Be systematic and work from one end of the fuel delivery system to the other.*
- *Loosen each union in turn and check to see whether there is a good flow of fuel from the open pipe before retightening.*
- *Using this method, you should be able to find the partially blocked section of pipe.*

SAFETY FIRST!
- *If you blow through pipes with compressed air, wear suitable eye protection.*
- *Make sure the open ends of any pipes are pointing away from you or anybody else in the area.*

Check 15. Blocked fuel tank vent.

See *Check 4*.

Check 16.
Incorrect valve clearances.

If the clearance between the end of the valve stem and the rocker arm is too small, the valves may not seat properly, which can result in poor starting and running because of:
• poor compression,
• a build up of deposits on the valve seat,
• or burning of the valve and seat.

Too large a clearance will prevent the valves from fully opening and causes 'valve clatter'. Some modern engines may be fitted with hydraulic valve lifters which cannot be adjusted - although they can fail - so this is worth checking before assuming there is a problem with valve clearances.

See *PART C, Job 16* for adjustment information.

Check 17.
Broken or weak valve springs.

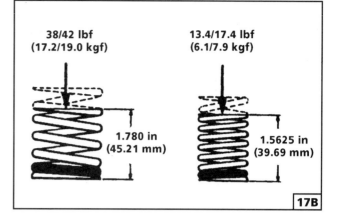

17B.

17B. Here are the figures for the springs fitted to a Perkins 6.3544 series engines. However, if you don't have access to the required facility and you check their uncompressed length against the manufacturer's recommended length, or the length of a new spring, you'll be able to tell if a replacement is needed. See *PART C*, for information on valve spring replacement.

Valve springs are reliable and have long lives. There is not much that can be done to predict or prevent weak or broken springs. Therefore, if the engine is being stripped down for any reason it is good practice, and recommended by manufacturers, to use new valve springs during the rebuild.

Check 18. Sticking valves.

Sticking valves can be caused by damaged valve stems or a build up of deposits on the valve stems but there is no way of telling the exact problem until the head is removed. Various chemical preparations are available which, when added to fuel or sprayed into the inlet manifold will, the manufacturers claim, clean all sorts of deposits from the valves, pistons and cylinders without the need for dismantling. WARNING: These preparations do work - but indiscriminately. They also remove deposits on pistons and rings which, on an older engine, can provide an extra, essential compression seal. **We do NOT recommend their use!**

With the cylinder head removed (see *PART C*) try to find out the cause of the problem. A bent or damaged valve may have been struck by the piston at some stage, while a build up of deposits may not be due only to worn valve guides. Look for other signs of wear such as worn or scored cylinder bores.

A low quality oil or elderly oil, overfuelling or general engine wear in other areas can cause a build up of deposits on valves which can result in sticking. Always use an oil suitable for the engine and its operating conditions, change it and the filter at the recommended intervals, and make sure the engine is adjusted and running correctly. Worn engines should be stripped down and overhauled.

17A. While broken valve springs are easy to spot, weak valve springs are not. If a valve spring has broken, there is a chance that the piston may have struck the valve causing damage to both components. A thorough check involves removal of the cylinder head and valve-gear stripdown.

i INSIDE INFORMATION: Weak valve springs can only be identified for sure by removing them and checking the force needed to compress them a set distance. *i*

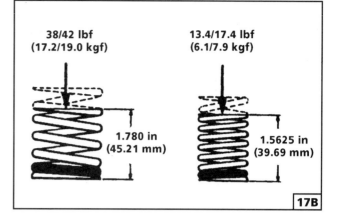
17A.

Check 19.
Incorrect injection pump timing.

Unless the fuel injection pump or the fuel injection pump drive has been disturbed it is unlikely that the fuel injection pump timing will be incorrect.

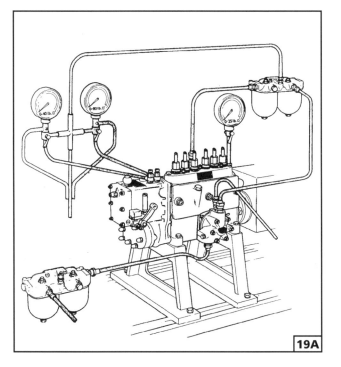

19A. If, after carrying out all the checks in the list, the engine still misfires and you suspect an incorrectly set or faulty injection pump, it must be checked with the appropriate test equipment.

19B. This is an example of the equipment used by Perkins to correctly set fuel injection pumps. The display on the screen gives the precise settings. Not exactly a back-yard job.

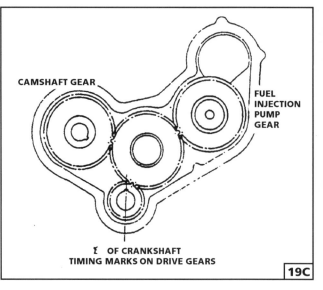

CAMSHAFT GEAR

FUEL INJECTION PUMP GEAR

Ⅰ OF CRANKSHAFT
TIMING MARKS ON DRIVE GEARS

19C. The manufacturer's manual will give details on the correct method for setting the camshaft and injection pump drive. Usually, irrespective of whether the drive is via belts, chains or gears, there are marks on the appropriate drive wheels which should be aligned with each other.

IMPORTANT NOTE: Turning the crankshaft quickly (such as by operating the starter motor) while the drive to the camshaft is disconnected, invariably results in the pistons striking the valves and irreparable damage. If it is necessary to turn the crankshaft, remove the glow plugs, turn the engine gently by hand and stop if any resistance is felt. Once you have reconnected the camshaft and injection pump drive, ensure there is no chance of damaging the valves and pistons. Turn the engine over at least twice by hand, stopping if any resistance is felt.

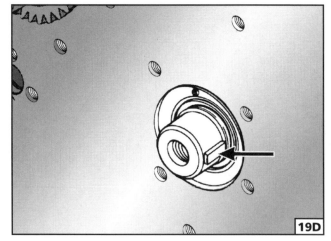

19D. There is a possibility, albeit remote, that the timing may be out because a drive wheel has rotated on its shaft. Drive wheels and pulleys often have keys which fit in slots on the shaft and keyways in the wheels. Commonly, the key is a half-moon (Woodruff) type (arrowed) although square types are used. If all else fails to solve the misfiring it's worth checking the condition - and presence! - of the key, just to make sure. Alternatively, some drive shafts have splined ends (described as serrated in some manuals) and these do not normally give problems, although there is more scope for incorrect assembly.

IMPORTANT NOTE: In normal running, it is unusual for the injection timing to suddenly become incorrect so, if it is 'out', it is more likely that the settings were wrongly set during reassembly or maintenance.

Check 20. Incorrect valve timing.

Since the injection pump is often driven from the camshaft, or from the same drive system, the information given in **Check 19** applies just as much to incorrect valve timing. Once again, there is unlikely to be a fault unless it was 'built-in' when the timing was last set.

Check 21.
Incorrect or faulty injectors

Much of what was said earlier about sticking injectors (see **Check 9**) also applies to faulty injectors.

Check the manufacturer's manual for the correct type of injector.

21. The CAV types shown here may have similar fittings but are not inter-changeable. Replace any that do not match the specification. Make sure that the correct sealing and insulating washers are fitted.

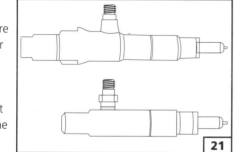

IMPORTANT NOTE: Make sure that no dirt has entered the injectors or any open fuel system pipes and that filters are in good condition.

SAFETY FIRST!

• *Diesel fuel is injected at very high pressures, so wear appropriate eye protection and do not let the fuel spurt or spray onto any part of the body.*
• *The pressures are high enough to force the fuel through the skin into the bloodstream with possibly fatal consequences.*

Check 22. Engine does not run at correct operating temperature.

See *FAULT 7* or *FAULT 8*.

Check 23.
Poor cylinder compressions.

See *Chapter 3, Part C*.

Check 24. Speed control linkages stiff or sticking.

There are many different designs of speed control linkage in use but most use mechanical linkages or cables. In general, the longer and more flexible the linkage, the greater the effect any stiffness or sticking will have. Larger input forces and movements have to be made to change the settings at the engine but these larger movements cause the engine speed to change more than was wanted, so the operator is constantly moving the controls but never getting exactly the right speed.

24. Have a good look at the linkage and see if there are any obvious causes of the stiffness or sticking. If the engine is remote from the control point, have somebody else move the control while you watch the linkage.

Here, a boat engine's speed control is some way from the engine.

If the problem is not obvious, disconnect part of the linkage and see what difference this makes. If there's no difference, the problem is on the control side of the discon-nection, while a noticeable difference indicates that the problem is on the engine side. Repeating this procedure closer to where you now think the linkage is sticking will normally identify the problem quite quickly.

Return springs can also present problems. Look for:
• too strong a spring fitted, as a replacement
• a spring that has been mounted in the wrong place, so that it restricts movement
• a spring that has been stretched and weakened
• a missing spring!

Check 25.
Faulty stop control operation.

Some engines have electric fuel cut-off solenoids while others have manual controls. On marine engines, the solenoid should be energised CLOSED but on land engines, such solenoids are energised ON. A common safety feature of industrial engines is an air shut-off valve.

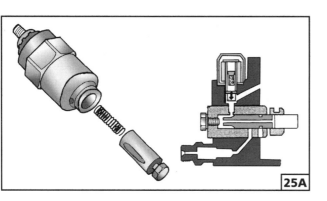

25A

25A. ELECTRIC SOLENOID: These are reliable and usually have long lives. Generally, when they give trouble it is a complete failure and the engine will simply stop, or continue running if a marine application.

• Use a multi-meter to check the resistance of the solenoid and the voltage across it under normal running conditions.
• The operation of manual stop controls should be checked to make sure that any linkage is correctly adjusted and that the cut-off moves through its full range without sticking.
• Check the operation of any emergency air shut-off valves to make sure they are operating when they should and are not sticking.

A very high solenoid coil resistance combined with measuring battery voltage across the terminal indicates an open circuit, while a very low resistance combined with a very low voltage drop indicates a short circuit. If the solenoid is serviceable, it is usually possible to feel it 'clunk' as it moves from the RUN to the STOP position. Solenoids can usually only be renewed, not repaired, unless there is a faulty terminal.

MECHANICAL AND EMERGENCY STOPS: Although most injection pumps do have mechanical stop controls, they are not often used as the normal stop control, but are backups in case of solenoid failure. Check the injection pump's mechanical override and make sure it is in the fully OPEN position.

All you can do with a mechanical stop system is check over the mechanism and make sure all the components are correctly adjusted and move through their full ranges and that the linkage is not stiff or sticking.

25B

25B. Check that any return springs are correctly fitted. An emergency stop system must be checked in accordance with the manufacturer's recommendations. Here the return spring is missing

Stop systems are usually reliable and trouble free. If they are kept clean and free from damp they should give few problems.

i INSIDE INFORMATION: In an emergency, if the speed control breaks, a mechanical stop control system can sometimes be used as a very basic engine speed control. *i*

Check 26. Choked fuel filter.

See **Check 5**.

Check 27. Faulty governor.

The governor is usually integral with the pump or mounted very close by. Mechanical and hydraulic governors are compact units but a pneumatic governor has linkages to the inlet manifold. Apart from making maximum and minimum speed adjustments, there is little that can be done with a faulty mechanical or hydraulic governor. Pneumatic governors can suffer from air leaks which can cause speed variations and surges. Electronic governors cannot be tested without sophisticated specialist equipment.

27A. MECHANICAL AND HYDRAULIC GOVERNOR: With a mechanical governor, the most you can do is check that the mechanism is free to move without sticking or binding. For both mechanical and hydraulic types you must consult the manufacturer's manual for precise setting up

27A

information and the correct test equipment must be used.

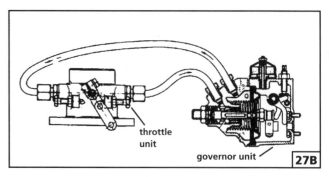

throttle unit

governor unit 27B

27B. PNEUMATIC GOVERNOR: There is a little more that can be done with pneumatic governors, the most import check being to make sure there are no air leaks in the pipes connecting the manifold venturi to the diaphragm or through the diaphragm itself.

A pneumatic governor will have at least one and possibly two air tubes connecting the inlet manifold venturi with the governor. You should check these to make sure they are not blocked. Watch out for tubes that have one way valves! Blow through in each direction (NOT with compressed air!) to check for correct operation.

27C. It is worth checking the mechanical linkages (**a**) from the speed control to the butterfly flap valve (**b**) in the venturi (**c**), and the flap valve itself. These mechanisms should operate smoothly without any sticking. Free off and lubricate as necessary.

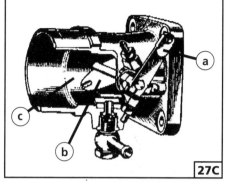

27D. If the pneumatic supply is okay and the linkages are free, the fault probably lies in the diaphragm (arrowed). Most diaphragms are made from neoprene rubber but you may encounter some made from leather. A stiff leather diaphragm can be oiled but one that is split must be renewed.

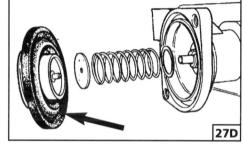

FAULT 3: ENGINE VIBRATES EXCESSIVELY

ℹ INSIDE INFORMATION: It is not uncommon for engine vibration to be mis-diagnosed as a misfire. If necessary, double-check or take a second opinion. **ℹ**

Check 28. Sticking or restricted speed control movement.

See **Check 24**.

Check 29. Faulty engine mountings.

The engine mounts and bed are subject to significant vibration in normal use and, in vehicle or marine applications, extra loads associated with vehicle movement. In marine applications the engine mounts can, in rough seas, be subjected to up to 6 times normal loads.

If an engine's vibrations are transmitted through the mountings to a vehicle or boat they can create discomfort for passengers, which is why most engines have flexible mountings. These mountings, usually made from special rubber bonded between two metal plates, are known as the 'Silentbloc' type. Rubber mountings can deteriorate with age and the rubber can separate from the metal. In some applications the flexible mounts may be metal springs which normally have very long lives.

If your engine is rigidly mounted, there is a possibility that part of the mounting structure may have broken because of vibration-induced metal fatigue.

29. Check the engine mountings while levering, lifting or trying to move the engine with a hoist - but not by so much that you damage the mountings or other components!

• A flexibly mounted engine should move a little but any excessive movement indicates a problem with the mounts.
• Rubber mounts may appear in good condition but can soften with age.
• The steel plates can become detached from the rubber - undetectable without physically levering or lifting the engine.

It is not normally possible to repair rubber or spring mounts and renewal is necessary. Use the correct type, as fitting mounts designed for other applications can sometimes make vibration worse. Rigid mountings can be repaired but consult the manufacturer for the correct techniques.

Check 30. Damaged or loose fan or crankshaft pulley.

COOLING FAN

30A. Look at the fan to see if there is any obvious damage, such as twisted or missing blades. But note that some modern fan designs have the blades unevenly spaced, which could create the impression that blades are missing.

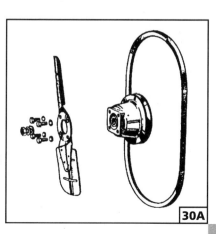

Try moving the fan and crankshaft pulley around to see if they are loose. If the fan drive belt is tight this may mask any looseness, so slacken it off before checking.

30B. If the fan is loose, it is usually a simple job to retighten the mounting bolts (arrowed) - although you may have to renew them if their threads are stripped because of chaffing. Tighten to the correct torque and use the correct lockwashers or other locking medium, as recommended in the manual.

30B

CRANKSHAFT PULLEY

Before tightening the crankshaft pulley, do check to make sure that the half-moon (Woodruff) key or spline, which holds it in place on the crankshaft, is not damaged or missing, by removing the pulley. Make sure the key is in place on the crankshaft and that the slot in the pulley has not been damaged or widened. Refit the pulley and tighten the retaining bolt or nut to the recommended torque. This is normally quite a high value, so you will have to lock the engine while you tighten the bolt or nut.

ℹ INSIDE INFORMATION: • Also, check for alternative crankshaft pulley locking devices, such as the 'Ringfeder' system. The method employs interlocking tapered rings and is used on several Perkins engines.
• A loose crankshaft pulley may have damaged the front oil seal, so check and renew if necessary.
• A faulty crankshaft pulley balancer unit, if fitted, can cause vibration. Balancers are timed and therefore, can be incorrectly timed. ℹ

Check 31. Wrong or incorrectly connected high pressure pipes.

See **Check 11**.

Check 32. Sticking valves.

See **Check 18**.

Check 33. Incorrectly aligned or loose flywheel.

These problems usually result from a lack of care during a recent rebuild. If you are lucky, you may be able to see the flywheel, or at least some part of it, without dismantling but this is unlikely.

• Try moving the flywheel in all directions. If you feel any movement other than rotation, or there is free movement before the flywheel rotates the crankshaft, the flywheel is loose.
• Any misalignment should be obvious but you can check more accurately by fixing a dial gauge (or a rigid pointer) onto a convenient part of the engine so it is just touching the flywheel.
• Take out the glow plugs, to remove compression and prevent the engine from starting, rotate the engine by hand, and see whether the flywheel runs true.

A loose or misaligned flywheel normally means the engine has to come out for repair on the bench or in a workshop. The actual realignment or tightening is relatively simple but there is almost certain to be collateral damage caused by the unaligned mass of the flywheel. Check especially (and renew if necessary) the end of the crankshaft, mounting bolts, flywheel housing, clutch components and the flywheel itself, especially at its mounting points. Follow the assembly instructions in the manufacturer's manual, especially those relating to torque loadings and the use of any thread locking compounds.

IMPORTANT NOTE: During any engine build, do check at every stage to make sure that components are assembled correctly and the correct torques are applied to nuts and bolts and that mounting bolts are of the correct length.

Check 34. Poor cylinder compressions.

See **Chapter 3, Part C**.

Check 35. Incorrect or faulty injectors.

See **Check 21**.

Check 36. Incorrect injection pump timing.

See **Check 19**.

Check 37.
Incorrect idle speed setting.

There are many different designs of diesel engine and the idle speed setting procedures will vary depending on type. You must consult the manufacturer's manual so that you can identify the controls used to set the idle speed, and be aware of any special precautions. Do not make adjustments at random in the hope that you get it right! At best you may upset other engine settings and at worst you may cause internal damage to components such as the injection fuel pump.

37A. If your engine has a tachometer or rev counter fitted, it is relatively simple to adjust the idle speed setting controls to achieve the correct speed.

On most engines, the idle speed is set by a simple screw adjuster (arrowed). On engines without tachometers you will have to use guesswork or, preferably, a hand held tachometer.

SAFETY FIRST!

• Hand held tachometers are used in close proximity to quickly rotating components. They should be used strictly as instructed by the manufacturer. Long hair and loose clothing should be kept well away from rotating machinery and drive belts.

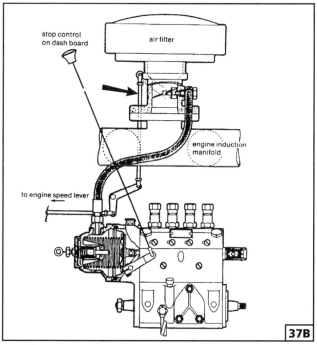

37B. If you have a pneumatic governor, it may also be necessary to adjust the butterfly valve operating mechanism (arrowed).

If components have been renewed or other changes made, these may have had an effect on the idle speed. However, the idle speed should not need frequent adjusting and if it does you should dig deeper to find out why.

Check 38. Leaking injector pipes (fuel escaping).

See **Check 3**.

Check 39.
Incorrect valve clearances.

See **Check 16** and **Check 18**.

FAULT 4: ENGINE KNOCKS EXCESSIVELY

Check 40.
Incorrect type or grade of fuel.

Continuing to run an excessively knocking engine can result in permanent damage, which can cost far more to repair than the cost of replacement fuel. But it is very difficult to tell, without chemical analysis, whether you have the wrong fuel in your tank.

Obtain a small quantity of fuel that you know to be the right type and grade for your engine. Now set up an emergency fuel supply to the injection pump, as described in **Chapter 3, Part C**. Start the engine and run it. If the fuel in the tank is the wrong type or grade, any excessive knocking should disappear, as good fuel flows to the injectors.

Once you're sure the fuel in the tank is wrong, the decision has to be made on how to get it out and where to put it. This will depend on the type and location of the tank and the quantity of fuel. You have the choice of pumping the fuel out, siphoning it or draining it. Pumping will work in most situations while siphoning and draining will only work if the receiving container is lower than the tank. Do not worry about getting every last drop of fuel out. If you completely refill the tank, any remaining old fuel will be well diluted and won't cause any harm.

IMPORTANT NOTE: With all the clatter of a diesel engine, it is often difficult to tell whether the engine is 'knocking' or not. Don't take chances! Always buy fuel from reputable suppliers and be sure you know which types and grades are suitable for your engine.

Check 41.
Incorrect or faulty injectors.

See **Check 21**.

Check 42. Incorrect injection pump or valve timing.

See *Check 19* and *Check 20*.

Check 43. Incorrect valve clearances.

See *Check 16*.

Check 44. Broken valve springs.

See *Check 17*.

Check 45. Sticking valves.

See *Chapter 3, Part C*.

Check 46. Incorrect piston height.

You are extremely unlikely to come across this fault in normal running, but it may arise in a recently rebuilt engine. Check the manufacturer's manual and make sure pistons are set to the height specified. If the pistons are set too low, the engine may be difficult to start and misfire during running.

It is sometimes possible, usually through carelessness, for water or other liquids to enter the intake manifold when an engine is running. The (incompressible) liquid will be sucked into a cylinder and trapped by the piston on the compression stroke. Even if the engine stops, the connecting rod may have been bent and the piston will then be lower in the bore.

Piston heights can only be accurately checked with the cylinder head removed. Cylinder head removal and piston projection measuring are described in *Chapter 3, Part C*.

To avoid the risk of water in the bores, make sure water cannot enter the induction system and don't try to hose down the engine or surrounding area while the engine is running.

Check 47. Worn or damaged small-end or big-end bearings.

SOUND ADVICE: Worn big-end bearings make a sound very similar to the characteristic diesel knock, while worn small-end bearings make more of a rattle. Unless there are other symptoms of wear, such as low oil pressure, it is usually impossible to confirm these problems without stripping the engine.

Check 48. Excessive camshaft bearing wear.

A mechanic who is especially familiar with a particular type of engine may be able to diagnose excessive camshaft wear by the noise from the engine. However, in most cases, this fault can only be found by stripping the engine and physically checking the camshaft and bearings.

Check 49. Broken, worn or sticking piston rings.

Broken, worn or sticking piston ring can result in low cylinder compressions. Confirming this problem will require some dismantling since low compressions can also be caused by worn or damaged pistons, badly seating or sticking valves and leaking head gaskets. See *Chapter 3, Part C*.

Check 50. Worn or damaged pistons.

See *Check 49*.

Check 51. Excessive timing gear backlash.

You are unlikely to come across this problem on an engine fitted with a belt-driven injection pump and camshaft, provided the belt is properly tensioned - **and there are no teeth missing**. On engines fitted with gear or chain injection pump and camshaft drives, gears and tensioners can wear and chains do stretch over time.

Watch the injection pump and camshaft drives while rotating the engine backwards and forwards by hand. Excessive backlash is undesirable and will cause variation in pump injection and valve opening times.

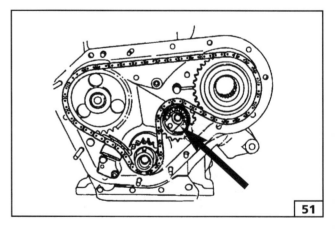

51. Chain type injection pump and camshaft drives have tensioners (arrowed) but these should not be used to take up the 'slack' caused by a worn, stretched chain. Renewal of worn gears and chains is the only solution. Follow the manufacturer's manual for information on the correct renewal and setting-up procedures.

Renew rubber belt drives at the periods specified by the manufacturer. Make sure that gear and chain drives are properly lubricated.

Check 52. Overfuelling through faulty or sticking governor or speed control system.

See **Check 27**.

Check 53. Faulty lift pump.

See **Check 12**.

Check 54. Faulty cold starting equipment.

The symptoms of this problem will only be apparent on start-up and will disappear once the engine is at running temperature. However, a cold start system that doesn't 'switch off' may cause excessive knocking when the engine is at normal operating temperature.

GLOW PLUGS

The most common type of cold start aid is the glow plug, an essential aid for indirect injection engines. Both the heating element and the duration of operation are electrically operated. As a result, problems may be due to the electric supply as much as to the glow plugs themselves. Although subjected to a very hostile environment inside the engine, glow plugs are quite fragile and can be easily damaged when removed.

It is unlikely that all glow plugs in an engine will fail at the same time so, if the engine shows no inclination to start at all, the problem is probably with the common supply. An engine that fires on a few cylinders before 'catching' completely may be symptomatic of some failed glow plugs.

i INSIDE INFORMATION: Beware of glow plugs that have a low resistance and operate on a pulsed supply. Connect them to a continuous supply, such as from the battery and they will burn out within seconds. *i*

54A. Glow plugs (**1**) are wired in parallel and power is supplied through a relay (solenoid switch-**3**). The location and type of relay will depend on the engine manufacturer and a wiring diagram for the particular engine should be used to identify and locate the relay and associated wiring. The power

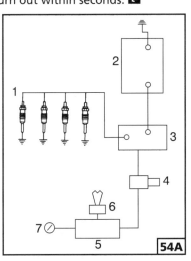

supply to the relay, timer and plugs should all be checked. Bear in mind that the glow plugs may only operate for a 'burst' of 20 seconds, so any checks will have to be made quickly.

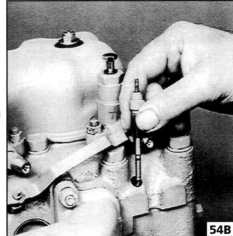

54B. There is not much you can do about a faulty glow plug, relay or timer, except renew it.

AIR PRE-HEATERS

You may come across cold starting aids that heat the air in the inlet manifold. Some of these are electric heaters similar to glow plugs. Other types, such as the CAV 'Thermostart' system burn fuel in the inlet manifold to pre-heat the intake air. Note that if this system continues to operate during normal running, excessive combustion chamber temperatures may be generated. Also, a fuel leak from the Thermostart system will cause over-fuelling.

• Disconnect the electrical supply to the Thermostart system. If this cures the problem, investigate and correct the fault with the electrical system.

• Disconnect the fuel supply to the Thermostart system (but obviously, not the supply to the injectors!). If this cures the poor starting problem there is a fault in the Thermostart unit which should be overhauled or renewed.

Check 55. Overheating.

See **FAULT 8: ENGINE OVERHEATS, Check 89** to **Check 101**.

Check 56. Piston seizure/pickup.

A piston that is starting to seize in the bore, or picking up on the cylinder walls creates a lot of friction and extra heat which can create uneven running and knocking. Stripping the engine down is the only way to find this problem.

FAULT 5: ENGINE WILL NOT ACCELERATE OR PRODUCE POWER

Check 57. Stop control not in fully ON position.

See *Check 25*.

Check 58. Engine speed control sticking or movement restricted.

See *Check 24*.

Check 59. Restriction in induction system.

See *Check 7*.

Check 60. Air in fuel system.

See *Check 2*.

Check 61. Partly blocked fuel feed pipe.

See *Check 14*.

Check 62. Choked fuel filters.

See *Check 5*.

Check 63. Blocked fuel tank vent.

See *Check 4*.

Check 64. Faulty lift pump.

See *Check 12*.

Check 65. Incorrect injection pump or valve timing.

See *Check 19* and *Check 20*.

Check 66. Restriction in exhaust system.

See *Check 8*.

Check 67. Faulty, worn or incorrect injectors.

See *Check 21*.

Check 68. Faulty, worn or incorrect injection pump.

Check the manufacturer's manual and make sure you have the right pump for the engine. If the pump is the wrong specification, it is very unlikely the engine will ever run properly. When fitting a replacement, follow the manufacturer's guidance and setting up instructions. Injection pumps are extremely reliable and, assuming the fuel is adequately filtered, will have very long lives. When they do go wrong there is not much you can do about it. The pump will have to be overhauled or renewed with a new one.

Remove the pump, taking note of the way any linkages are connected, and any shims that are used to adjust the pump timing. All openings must be sealed to prevent dirt getting into the pump.

When the new or overhauled pump is fitted, it will have to be accurately timed, using the correct equipment and following the manufacturer's instructions.

Check 69. Incorrect valve clearances.

See *Check 16*.

Check 70. Poor cylinder compressions.

See *Chapter 3, Part C*.

Check 71. Poor boost pressure (turbocharged engines).

Specialist test equipment is normally needed to properly diagnose turbocharger problems. If a boost gauge is fitted, compare the readings with the manufacturer's recommendations. However, before assuming that the turbocharger is fundamentally faulty, check to make sure that the turbo. wastegate is not jammed open.

Check 72. Sticking fuel delivery valves.

Fuel injection pumps have delivery valves at their high-pressure outlets. These valves control the rise and fall of pressure in the high-pressure fuel lines to the injectors. Any problems with fuel delivery valves can affect the operation of the injectors. For example a valve that sticks open will result in a more gradual rate of pressure rise and fall at the injector which can lead to premature injection and chatter.

FAULT 6: EXCESSIVE LUBRICATING OIL CONSUMPTION

Check 73. Oil leak.

Most modern engines deliver oil through galleries and drillings within the engine and usually, the only external oil pipes are those to oil coolers. The most common places for oil to leak are at oil pipe unions and joints between engine components, although on older-design engines, there may be other external pipes to check.

Look carefully at the engine and underneath it to see if any oil leaks are obvious. Don't forget to look at the coolant (allow it to cool down), where brown globules or off-white 'salad cream' indicate a leaking head gasket. A pool of oil near the filler might just be a spill, so don't jump to conclusions!

Check 74. Oil level too high.

A high oil level through over-filling is usually a result of checking the oil level just after the engine has run. Oil in circulation may not have had time to drain back to the sump, so the oil level looks low and is topped up. A high oil level may indicate a fuel leak into the sump, so check that out as well.

Using the sump drain plug to remove a small quantity of oil doesn't work. You'll need to use a small suction pump, usually through the dipstick hole.

Make sure the engine has been standing for some time on a level surface. Drain small quantities at a time and check the level. As a guide, the difference between the MAX and MIN marks on most small engine dipsticks is between half-a-litre and one litre, depending on model. Make sure any probe that goes into the sump is clean and free from dirt or grit.

i INSIDE INFORMATION: Make sure that the stop on the dipstick has not moved (not unknown!) and that the dipstick is the correct one for the engine, otherwise you won't be able to judge the correct level. *i*

Check 75. Engine breathing system blocked.

Even new piston rings 'leak'. This leakage causes a build up of pressure in the crankcase which, if not vented, escapes anywhere it can, taking a quantity of oil with it. It follows, therefore, that a badly worn engine will create excessive fumes and, in *extremis*, this can cause a blockage.

In the past, excess pressure was vented to the atmosphere, but these days we are - quite rightly - more environmentally aware, and this leakage, which contains all sorts of unpleasant compounds, is either recirculated through the inlet manifold or injected into the exhaust.

Breather systems don't often get blocked and if they're cleaned out routinely, they shouldn't create any problems. You will need to check with the manufacturer to see if any pumps and valves which are found to be faulty are repairable.

First, find out what type of breather system is used and where all the components and pipes are located. Some systems run the crankcase at atmospheric pressure while others, using scavenge pumps, slightly pressurises the crankcase or use a vacuum to remove the undesirable combustion by-products.

Check each part of the system until the blockage is found. Clean out the pipes or renew damaged or collapsed pipes and reassemble the system. Make sure any pumps or valves are working.

Check 76. Worn valve stems or valve guide bores.

Worn valve stems and valve guide bores will allow oil to seep into the inlet or exhaust manifolds. Excessive oil consumption, usually combined with blue smoke from the exhaust is an indication of oil burning in the combustion chamber or exhaust manifold. However, these symptoms have other causes so dismantling is the only way to confirm that valve stem or valve guide wear is causing the problem. See **Chapter 3, Part C**.

Check 77. Worn valve stem seals.

Worn valve stem seals, especially when combined with worn valve stems or valve guide bores will allow oil to seep into the manifolds. Valve stem seals can normally be replaced with the cylinder head in situ but if the seals are worn, other wear may be contributing to the problem. Without dismantling the cylinder head it is difficult to tell whether the valve stems and valve guide bores are also worn.

See **Chapter 3, Part C**.

Check 78. Worn cylinder bores or pistons.

See **Check 50**.

Check 79 Worn or broken piston rings.

See **Check 49**.

Check 80. Wrong type of oil, diluted oil or inferior quality oil.

Use of the incorrect grade of lubricating oil is a possible cause of excessive consumption. However, without an expensive chemical analysis, there is no definite way of checking the type of oil in the sump and a badly worn engine is a more probable cause of the excessive consumption. If you are convinced the engine is in good condition, changing the oil and filter is the easiest way of making sure the correct grade of lubricant is in the sump.

Check 81. New or rebuilt engine not fully bedded in.

Until pistons, rings and bores bed in, a new or rebuilt engine will use more oil than one that has been fully run-in. Follow the manufacturer's running-in instructions and change the oil and filters at the more frequent periods specified. Although modern material and lubricants do not make the running-in process as critical as it used to be, an engine will benefit from being properly run-in.

Check 82. Glazed cylinder bores.

Engines start out their lives with slightly rough honed cylinder bores, which retains oil to lubricate the pistons and piston rings. Over many operating hours, this rough surface is both worn smooth and coated with gums and varnishes produced by the combustion process. This smooth, glazed surface allows more oil to escape past the piston rings which results in higher oil consumption than with the original, slightly rough, honed surface.

If glazed bores are suspected as the cause of high oil consumption, the engine will have to be stripped down. See *Chapter 3, Part C*.

Check 83. Faulty oil cooler.

Oil coolers can be air-or liquid-cooled. The air-cooled type is like the familiar car radiator, with air passing through a matrix of oil filled tubes. Liquid-cooled types transfer the heat in the oil to another liquid rather than air. Oil leaking out of an air-cooled type is easy to spot but might be more difficult with liquid-cooled types.

Look at the exterior of the cooler and see if there are any obvious leaks. It is possible for the liquid-cooled types to leak oil internally into the cooling liquid, so with this type, check the coolant as well.

It is sometimes possible to repair a faulty oil cooler with solder but renewal is preferable if parts are available.

Check 84. Cross leakage between oil feed pipe and fuel pipe.

If oil and fuel pipes are pressed very hard against each other they may chafe and wear, allowing oil to pass between them. A leak of this nature is extremely unusual, however.

• Follow the route of the fuel pipes and any external oil pipes. Separate any that are pressing hard against each other.

• Run the engine and see if oil or fuel leaks from the pipes. Renewal is the only practical solution

• In the longer term, make sure that any existing or new oil pipes do not press hard against fuel pipes.

Check 85. Oil leaks from ancillary equipment.

Some ancillaries may be lubricated by engine oils while - conversely - others may have seals or gaskets which prevent engine oil passing into them. If enough oil is being carried over for you to notice excessive oil consumption, there is a good chance that you will notice oil leaking from the ancillary component.

Inspect all ancillaries for any obvious oil carry-over such as drips, dribbles or runs.

Check 86. Consumption by the fuel injection pump.

Some injection pumps, particularly in-line types, are partly lubricated by engine oil. Wear in the pump may allow the oil to leak into the section lubricated by the diesel oil. If the wear is this bad, the pump really needs to be overhauled or renewed. See *Job 68*.

FAULT 7: ENGINE RUNS COLD

Check 87. No thermostat, faulty thermostat, or wrong thermostat fitted.

If an engine runs at a lower temperature than intended, it will not operate as well as it should. Certain parts may wear prematurely because they are not at their ideal working clearances, the oil may sludge prematurely and, in general, the engine will operate at less than its maximum thermal efficiency.

87. Remove the thermostat housing and check to see if a thermostat, or thermostats, are fitted. Some engines, such as the Perkins 12 and 16 cylinder SE vee types have a thermostat for each bank of cylinders while some others, such as the Perkins C and SF 65C ranges have several thermostat capsules in one unit as shown here.

87

If a thermostat is there, remove it and check the specification. Various numbers are often stamped on thermostats and one of these will be the opening temperature. The manufacturer's manual should tell you the intended opening temperature, which may be different for summer and winter. Test the thermostat by suspending it in a pan of warm water along with a suitable thermometer. Heat the water up and note the temperature when the thermostat starts to open. This should be within a few degrees of the figure stamped on it.

If the thermostat is working, refit it using a new gasket to seal the thermostat housing. An unserviceable thermostat should be thrown away and a new one fitted. A missing thermostat should, of course, be replaced with a new one. Thermostats are reliable and do not often give trouble. When they go wrong they cannot be repaired and have to be renewed.

Check 88. Faulty gauge or temperature transmitter.

Unless you have access to the figures used by manufacturers to check their transmitters and gauges, there are very few checks that can be carried out. The checks that can be carried out are often made more difficult when gauges are marked with the words COLD, NORMAL and HOT or simply with blue, green and red coloured sectors. However, the thermostat opening temperature can be taken as a fair guide to the operating temperature.

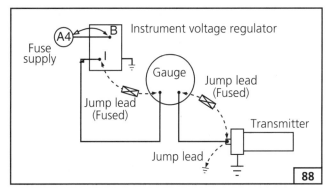

88. Start by checking the electrical circuit for continuity, or the capillary tube (from sender bulb to gauge) for crushing, kinks or splits, depending on type.

ELECTRICAL SYSTEMS

• A fault could be in the electrical gauge, transmitter or wiring, so all three areas need to be checked. Do make sure that you are testing the right circuit, since modern engines have many similar looking transmitters and it is easy to confuse them.

• Check the electrical gauge by open circuiting the circuit at the transmitter, then shorting the connection to earth/ground. Normally, the gauge will show zero when open circuited and maximum when shorted. If the gauge responds, the gauge and wiring should be okay.

• If the electrics (or the tubing) are okay, remove the transmitter from the engine and reconnect it to the gauge. (An old-style transmitter containing a liquid - i.e. those gauges

with a copper tube - cannot be disconnected from the gauge.) Heat up a pan of water to boiling point and put the transmitter into the water along with a suitable thermometer. When the reading on the thermometer is at the thermostat's opening temperature, look at the reading on the gauge. If the reading is well below the green, mid or NORMAL sections there is a fault with the gauge or transmitter.

FAULT 8: ENGINE OVERHEATS

Check 89. Insufficient coolant.

A diesel engine can covert between 30% and 40% of the heat in the fuel into useful work. The other 60% to 70% has to be got rid of, otherwise the engine will overheat and seize. So it is vital that coolant levels are maintained and the reasons for any losses are found and put right. Check the level of the coolant in the radiator or header tank when the engine is cold, and top up as necessary.

SAFETY FIRST!

• *Do not attempt to remove the filler cap or dismantle any part of the cooling system when the engine is hot. Engine cooling systems are normally pressurised, which prevents the coolant boiling until a higher temperature than if it was at atmospheric pressure. Depressurising the system can allow the coolant to boil, with steam and boiling liquid spraying out of the opening at considerable force.*

Look at all the hoses and hose clips to see if there are any obvious leaks. Leaking clips can be tightened but don't overdo it. Metal corrosion on the pipe outlet may be preventing a good seal - remove and check, if necessary.

Feel all the hoses to see if any have gone soft, or for signs of internal deterioration, usually evidenced from a crackling feeling inside the pipe when squeezed by hand. Don't forget also to check the radiator condition. If there are no obvious external leaks, the possibility of internal leaks, possibly due to gasket failure may be considered.

89. In *extremis*, a pipe stub can corrode to this extent! Look for evidence of leaks (which will invariably precede this stage) and remove hoses for checking, if tightening doesn't make any difference.

Check 90. Oil level too high.

If the oil level in the sump is too high, the crankshaft and conrods will churn the oil as they rotate. This can generate extra heat, which may be more than the cooling system can dissipate. It could also encourage oil to be forced up the bores, leading to poor running.

See **Check 74**.

Check 91. Faulty thermostat.

See **Check 87** - but for the obverse of a thermostat stuck fully open: one which is jammed shut.

Check 92. Blocked coolant system, faulty or incorrect radiator, coolant hoses or pressure cap.

Unless you have renewed any parts of the cooling system recently, it's unlikely that the wrong components are fitted but it's always worth checking. Rubber hoses do deteriorate over the years and radiator caps can corrode. Sludge can build up in coolant systems, eventually blocking the narrow passages in radiators, especially if there is no anti-freeze in the coolant or the anti-freeze is not changed regularly. Anti-freeze contains inhibitors which protect the engine from internal corrosion and the formation of sludge.

When you are sure that all the hoses and components are in good condition, flush the coolant system using one of the preparations available for this job. Follow the instructions carefully, paying particular attention to any safety warnings.

Sometimes flushing from the 'top down' doesn't clear a blockage and the system needs to be back-flushed. Here, the flushing water is forced through the system in the direction opposite to the normal coolant flow: in a conventional radiator system the water would be forced in through the 'cold' outlet from the radiator, from bottom to top.

Check 93. Loose fan belt.

The fan belt will usually tell you if it's loose by making a loud squeal when the engine runs - especially when it is accelerated. Tightening it up is an easy job but be careful not to make it too tight. Overtightening puts very heavy loads on water pump and alternator bearings, causing premature wear.

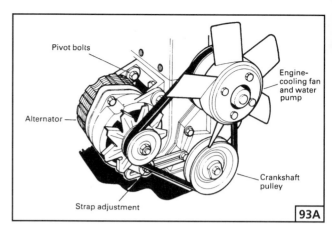

93A. On most engines, fan belt tension is controlled by adjusting the alternator or generator mountings but do check your engine manufacturer's manual for any special requirements.

Most vee-shaped fan belts require 12.5 mm (1/2 in.) of free play on the longest section, but check the manufacturer's manual.

93B. Before tightening, check the inside of a vee-belt for cracking, fraying or a glazed surface. If any of these are found, the belt is ripe for renewal.

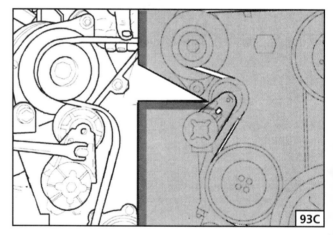

93C. Some modern engines use a flat belt with several small 'vees' which looks like a cross between a normal vee-belt and a toothed belt. Although the consequences of fan and alternator drive belt failure are less disastrous than camshaft drive belt failure, this flat type of belt should be handled with as much care as a toothed belt and the same precautions taken. You should consult the manufacturer's manual for information on belt replacement and tensioning. For example, on the Perkins Peregrine and 1300 Series engines this belt is kept at the correct tension by a spring loaded idler wheel.

Check 94. Faulty water pump.

Water pumps are simple devices and the most common problems will be coolant leakage or worn bearings. Leaking coolant should be quite obvious and worn bearings will either be noisy or there will be free play in the drive shaft. A tight fan belt will hide the free play, so loosen the fan belt before checking. Most modern water pumps are not repairable but it is worth checking with the manufacturer to see if repair kits are available.

making it easy • *Look out for a brown (or anti-freeze coloured) coolant stain on the engine block, beneath the water pump - a good clue that failure may be imminent.*

Some older water pumps have lubrication points, so check the manufacturer's manual for servicing requirements. Most modern water pumps are sealed for life so there is not much to do until they go wrong. Whichever type you have, check that the drive belt is not too tight. An overtight drive belt will cause premature wear of the pump bearings.

Check 95. Leaking cylinder head gasket or cracked cylinder head.

Serious overheating, topping up the coolant with cold water while the engine is hot or allowing the coolant to freeze in cold weather can cause cylinder head cracks. Unless the crack is very large, it will be difficult to detect and may require specialist examination. If you know that the engine has suffered any of the abuse listed here, it would be worth having the cylinder head crack tested. See **Job 95** for information on cylinder head gasket problems.

Detecting this fault will depend on where the crack or gasket failure is located.

The most obvious evidence of head gasket leakage or a cracked cylinder head is bubbles of gas in the coolant. Remove the pressure cap (ONLY with the engine cold) and run the engine. If any bubbles of gas are present (only apparent after the thermostat has opened, allowing a flow through the radiator), and especially if accompanied by oil in the coolant, it is fairly certain you have head gasket problems or worse, a cracked cylinder head.

Remove the cylinder head and examine the remains of the gasket. Sometimes gasket failure is obvious but not always.

If the gasket does not appear to have been leaking, have the cylinder head crack tested. See **Job 95**.

Check 96. Incorrect injection pump or valve timing.

See **Check 19** and **Check 20**.

Check 97. Faulty or incorrect injectors.

See **Check 21**.

Check 98. Restriction in induction system.

See **Check 7**.

Check 99. Restriction in exhaust system.

See **Check 8**.

Check 100. Blocked gearbox or engine oil cooler.

Not all engines have gearbox or engine oil coolers, so check the system or manufacturer's manual to see if one is fitted. You are very unlikely to come across this problem with modern oils, changed at the recommended periods. However, it is possible for debris such as bits of cleaning rag or paper wipe to enter the engine and be picked up by the oil. Before dismantling anything, feel the temperature of the engine at the oil outlet to the cooler and again at the cooler, when the engine has warmed up to normal operating temperature. Watch out, the engine may be very hot! A big difference between the two probably means no flow to the cooler - although if an oil cooler thermostat is fitted, it could be that the engine oil is not hot enough for oil to be diverted to the cooler.

Disconnect and remove the cooler. Unless the blockage is very obvious, such as in the supply or return pipes or right at the inlet or outlet to the cooler, it is best to hand the cooler to a specialist for flushing and unblocking.

Check out and, if necessary, flush the rest of the lubrication system before refitting the cooler.

Check 101. Faulty gauge or temperature transmitter.

See **Check 88**.

FAULT 9: OIL PRESSURE TOO LOW

Check 102. Engine lubricating oil too thin.

See **Check 80**.

Check 103.
Worn or damaged bearings.

Worn bearings are unlikely on a new or newish engine but damage can happen at any time. A knocking noise, similar to the typical diesel knock, means that the big-end bearings are worn. However, the noise of a diesel engine can make big end knock difficult to identify.

Damage is usually due to debris by-passing the filter and scoring the surfaces of the bearings. Oil filters have by-pass valves which will open and allow unfiltered oil to the engine if the filter element becomes blocked, so change the filter at the recommended times! Using a thicker oil may improve the oil pressure but, since you don't know whether the low oil pressure is due to wear or damage, it is best to overhaul the engine. See **Chapter 3, Part C**.

Check 104.
Insufficient oil in sump.

If you know that your engine burns oil or has a leak, check the oil at frequent intervals. Although the oil's primary job is lubrication, it also cools the engine, and especially the pistons. Fix any significant oil leaks and, if your engine is burning oil, carry out an overhaul.

Check 105. Faulty gauge or pressure transmitter.

Check the electrical circuit between the transmitter and gauge. On older systems that transmit pressure to the gauge through a capillary tube, make sure the tube (and sender) is in good condition and not blocked, holed, or kinked.

105. Remove the oil pressure transmitter and fit a reliable pressure gauge in its place. Use of an adapter is usually necessary. Make sure the gauge you use is suitable for the oil pressure range being measured. Run the engine and compare the pressure readings with those in the manufacturer's manual.

Unless you have the equipment and calibration data, once you have confirmed that the oil pressure is satisfactory, all you can do is renew components one at a time until the oil pressure reads correctly.

Assuming the test shows the correct pressure, try renewing the transmitter. If the pressure shown on the gauge is still low, you can renew the gauge or try dismantling it to see if the pointer position can be adjusted.

Check 106. Worn oil pump.

Oil pumps do not normally receive the benefits of filtered oil and any dirt or debris in the oil can cause quite rapid wear. Unfortunately, dismantling is necessary before any checks can be carried out. See **Chapter 3, Part C**.

Check 107.
Pressure relief valve stuck open.

Some oil pressure relief valves are easily accessible while others are in the bowels of the engine, beside the oil pump. Check the manufacturer's manual for the location. Any sticking is usually due to debris in the oil and the very act of dismantling the valve often cures the problem.

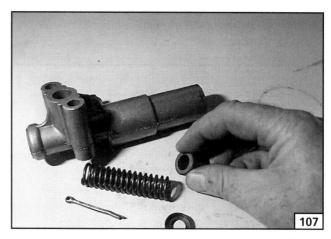

107. Dismantle the oil pressure relief valve and look for any obvious signs of dirt or debris. Pay particular attention to any scoring or pitting of the valve seat. Renew any damaged components. Reassemble and refit the relief valve. Run the engine and check the oil pressure.

Check 108.
Broken relief valve spring.

See **Check 107**.

Check 109. Faulty suction pipe.

See **Job 109**.

Check 110. Blocked oil filter.

See **Job 110**.

Check 111. Gearbox or engine oil cooler blocked.

See *Check 100*.

Check 112. Blocked sump strainer.

The sump strainer is a coarse filter which prevents large piece of debris getting to the oil filter and pressure relief valve. It will only be blocked if there is gross contamination of the oil in the sump.

FAULT 10:
OIL PRESSURE TOO HIGH

Check 113.
Engine lubricating oil too thick.

See *Check 80*.

Check 114. Faulty gauge or pressure transmitter.

See *Check 105*.

Check 115.
Pressure relief valve sticking

See *Check 107*.

FAULT 11:
ENGINE OVERSPEEDS

Check 116.
Incorrect maximum speed setting.

There are many different designs of maximum speed setting procedures. You should consult the manufacturer's manual, so that you can identify the controls used to set the maximum speed and be aware of any special precautions. Do not make adjustments at random in the hope that you get it right! the maximum speed setting is critical and if set too high can result in the engine becoming severely damaged.

IMPORTANT NOTE: The most recent engines are liable to emissions laws which forbid the adjustment of the maximum speed setting unless the adjustment is carried out by an approved person.

If your engine has a reliable tachometer or rev counter fitted it is relatively simple to adjust the maximum speed setting controls to achieve the correct speed. On engines without tachometers you will have to use a hand held tachometer.

SAFETY FIRST!

• *Hand held tachometers are used in close proximity to quickly rotating components. They should be used strictly as instructed by the manufacturer. Long hair and loose clothing should be kept well away from rotating machinery and drive belts.*

On most engines, the maximum speed is set by a simple screw adjuster. This is likely to be close to the idle speed setting screw so make sure you know which is which before making any adjustments. If you have a pneumatic governor, it may also be necessary to make adjustments to the venturi valve.

Component renewal or other changes may affect the maximum speed. However, the maximum speed should not need frequent adjusting and if it does, you should dig deeper to identify and remedy the cause.

Check 117.
Faulty or incorrectly set governor.

See *Check 27*.

Check 118. Faulty or incorrectly set injection pump.

See *Check 68*.

FAULT 12:
ENGINE WILL NOT STOP

Check 119. Faulty, damaged or incorrectly set stop mechanism.

See *Check 25*.

Check 120. Fuel or oil leak into induction system or cylinders.

Diesel engines can run on a number of different fuels, including lubricating oils and the oil used in some types of air filter. If the engine's oil consumption is excessive, and it is producing a lot of blue or white smoke at the exhaust, lubricating oil is probably entering the cylinders, either past the piston rings or down through the valve guides. Smoke on its own, without excessive oil consumption, can mean that oil from an air filter is being sucked into the engine.

If the engine has an oil bath or oiled mesh air cleaner, remove the cleaner and see if this makes a difference. Clean the old oil out of the air cleaner and renew it (NOT over-filling it!) using the manufacturer's recommended procedures.

120

120. This cutaway gives some idea of the internals of an oil bath air cleaner

Oil-leaks past the piston rings, or down past valve guides, can only be cured by an engine overhaul. See *Chapter 3, Part C*.

SAFETY FIRST!

• A diesel engine will stop if either the fuel or air supply is blocked. In an emergency, if the engine will not stop, you can try stuffing a rag into the air inlet. If the engine revs. continue to rise uncontrollably and to a dangerous degree, clear all personnel from the vicinity before the engine self-destructs.

FAULT 13: EXCESSIVE FUEL CONSUMPTION

Check 121. Fuel leaks.

See *Check 3*.

Check 122. Restriction in induction system.

See *Check 7*.

Check 123. Sticking valves.

See *Check 18*.

Check 124. Incorrect valve clearances.

See *Check 16*.

Check 125. Faulty cold starting aid.

See *Check 54*.

Check 126. Incorrect injection pump or valve timing.

See *Check 19* and Check 20.

Check 127. Faulty or incorrect injectors.

See *Check 21*.

Check 128. Faulty or incorrect injection pump.

See *Check 68*.

Check 129. Poor cylinder compressions.

Piston ring and cylinder bore wear, or valves that are not closing or seating properly, are the most common causes of low compressions. However, do check that the engine is not fitted with a decompressor and if so, and that you are not leaving it open once the engine is up to a good cranking speed.

Piston ring and cylinder bore wear is normally accompanied by high lubricating oil consumption and blue smoke out of the exhaust. Unfortunately, identifying the exact cause and curing the problem invariably means a strip-down. See *PART C*.

A compression test will confirm whether the compression in all cylinders is low. See *Chapter 3, Part C.*

SAFETY FIRST!

• Use only a compression tester designed for diesel engines - remember that the compression pressures are much higher than those of petrol engines.
• Make sure the fuel supply is disconnected - any fuel in the cylinder may ignite and destroy the compression tester and cause injury as the tester disintegrates.
• Do not squirt oil into the cylinders to 'seal' the bores. Once again, the oil may ignite as the engine is turned over.

IMPORTANT NOTE: Not every Job shown in PART A and PART B is covered here. In other words, you will find some Jobs missing. The remedy in some cases is obvious (putting fuel in an empty tank for instance), while in others, the act of carrying out the check is the same as carrying out the repair, such as finding and clearing a blocked fuel line.

Every Job in this section has the same number as the relevant Check in PART B. In other words, where **Check 19** described how to diagnose "Incorrect injection pump timing". **Job 19** tells you how to carry out the necessary repairs, or in this case adjustments.

FAULT 1: ENGINE STARTS AND STOPS

Job 1. Insufficient fuel in tank.

See **Job 2** for bleeding and priming the fuel system.

Job 2. Air in the fuel system.

BLEEDING THE FUEL SYSTEM

If air enters the fuel system, if must be eliminated from the system before the engine can be started.

Air can enter the system if:
• The fuel tank is drained during normal operation.
• The low-pressure fuel pipes are disconnected.
• A part of the low-pressure fuel system leaks during engine operation.

IMPORTANT NOTE: Check to see whether you have a self-priming pump. If you undo bolts or screws randomly in a vain search for a bleed screw there is a good chance you will loosen some that hold vital internal components in place. At the very least, your injection pump will be out of adjustment but it's more likely that irreparable damage will be caused when you try to start the engine. Similarly, if you do not have a self priming pump make sure you know exactly where the bleed points are.

In order to remove air from the fuel system:

❑ **Step 1A.** Loosen the vent plug (arrowed) on top of the twin element fuel filter by two or three turns.

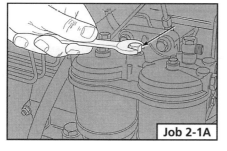
Job 2-1A

❑ **Step 1B.** If a single element filter is used, loosen the banjo connection bolt (arrowed) which is fitted to the top of the filter.

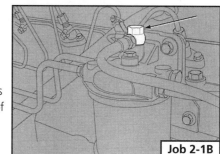
Job 2-1B

❑ **Step 2.** If the engine has a lift pump with a hand primer, operate the priming lever on the fuel lift pump (arrowed) until fuel, free of air, comes from the filter vent point. Tighten the vent plug or the banjo connection bolt.

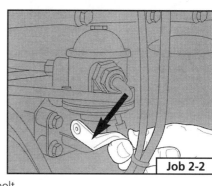
Job 2-2

ℹ INSIDE INFORMATION: If the drive cam of the fuel lift pump is at the point of maximum cam lift, it will not be possible to operate the priming lever. In this situation, the crankshaft must be rotated one revolution. ℹ

❑ **Step 3.** Ensure that the manual stop is in the 'RUN' position. If an electrical stop control is used, turn the key of the start switch to the 'START' or 'RUN' position.

❑ **Step 4.** Systematically bleed the fuel system. Start from the fuel tank and work towards the engine. Loosen each union in turn and allow fuel to run out until there is no trace of air. Retighten the unions but don't overdo it - overtightening unions can split them.

❑ **Step 5.** If a mechanical lift pump is fitted at the engine, loosen the union nut of the fuel inlet pipe (arrowed). Operate the priming lever of the fuel lift pump until fuel, free of air, comes

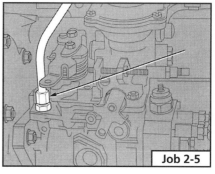
Job 2-5

from the loose connection. Tighten the union nut.

❑ **Step 6.** Loosen the union nut (arrowed) at the fuelled starting aid, if one is fitted, and operate the priming lever of the fuel lift pump until fuel, free of air, comes from

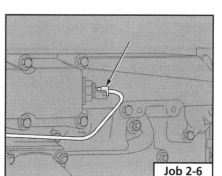
Job 2-6

the connection. Tighten the union nut at the starting aid.

☐ **Step 7.** Loosen the high-pressure pipe connections (arrowed) at two of the injectors.

Job 2-7

☐ **Step 8.** Ensure that the manual stop control, if one is fitted, is in the RUN position. Operate the starter motor until fuel, free from air, comes from the pipe connections.

☐ **Step 9.** The engine is now ready to start.

IMPORTANT NOTE: If the engine runs correctly for a short time and then stops or runs roughly, check again for air in the fuel system. If there is still air in the fuel system, there is probably a leakage in the suction or low-pressure system.

Job 3. Fuel system leaking.

☐ **Step 1.** Examine the system and look for clear evidence of fresh fuel running or dripping from pipes, unions or components. Internal leaks, such as from fuel pumps will be more difficult to spot. Check the engine oil for any fuel contamination.

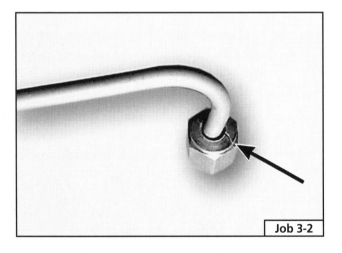

Job 3-2

☐ **Step 2.** Leaking unions can often be tightened but be careful! Overtightening unions can break them or strip threads, and further tightening only makes the situation worse. Look closely at the union to see if there are any obvious cracks and make sure it's square and not cross-threaded. Cracked or thread-stripped unions will have to be renewed: easy if the union is on a removable pipe but much more difficult on a component, unless the threaded sub-assembly can be removed and renewed, or the thread chased with the correct size of thread die.

☐ **Step 3.** Leaking pipes generally have to be renewed. Temporary repairs can sometimes be carried out to low pressure pipes using special compounds and, in desperation

(while in a dire emergency, such as at sea when there is nothing else available), chewing gum can be tried. In both cases, the pipe surface must be clean and fuel-free. Do not use these methods with high pressure pipes; always renew them.

☐ **Step 4.** External leaks from components such as pumps should be obvious. Internal leaks into the engine are more difficult to spot and even more difficult to trace to the offending component. If an external leak is found, renewal or overhaul is normally the only option but do check with the manufacturer for any repairs that can be carried out.

SAFETY FIRST!

• *Any fuel leaking from the high pressure delivery will have sufficient speed to penetrate skin with possibly lethal consequences, so do not allow any fuel to spray onto the body.*

• *Although diesel fuel is not as highly flammable as other fuels, the fine spray from a high pressure leak can be ignited quite easily.*

IMPORTANT NOTE: Overhaul any faulty components, following manufacturer's instructions, and paying particular attention to the cleanliness. Remember: dirt is death to many diesel engine components.

Job 7. Restriction in induction system.

Apart from a blocked air filter, a restriction in the induction system is unlikely unless of course the engine has an emergency air shut-off valve.

☐ **Step 1.** If the engine has an emergency air shut-off valve, check to see whether it is closed. Some types, e.g. those manufactured by Chalwyn Equipment and designed to protect against engine overspeeds, will reset automatically but others may need to be manually reset. If necessary, reset the valve by moving the manual reset lever from the 'CLOSED' to the 'OPEN' position.

SAFETY FIRST!

• *Find out why the air shut-off valve was triggered. Trying to restart the engine without finding the cause of shut-off could be dangerous or cause damage.*

☐ **Step 2.** Once you are certain there is no risk to safety, and if the air shut-off valve is open (or your engine does not have one), disconnect the air inlet to the manifold and try starting the engine. If the engine starts, you have a clogged air filter or an obstruction in the air inlet ducting. The only sure way of eliminating the filter as a cause of the intake restriction is to renew it.

FACT FILE

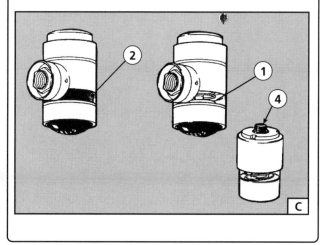

• Environmental conditions have an important effect on the frequency at which air cleaners and filters need to be serviced or renewed. You should bear this in mind when assessing potential faults and it is a good idea to get some idea from the operators about the conditions in which an engine runs.

• **A.** Some air filters have separate dust bowls (**1**) which must be cleaned before dust completely fills the bowl. Too much dust in the bowl and the life of the filter element is reduced.

• **B.** Other types of filters have automatic dust valves (**1**) to expel dust from the filter. If this doesn't work properly, dust builds up in the filter.

• **C.** Not so common, a restriction indicator is fitted to some inlet systems between the air filter outlet and manifold. A red warning indicator (**2**) shows when the pressure drop through the filter is too high i.e. the filter is clogged. After a clean element has been fitted press the rubber bottom or the button (**4**) on the restriction indicator to reset the red warning indicator (**1**).

Job 9. Sticking injectors.

❏ **Step 1A.** Injectors can be tested and cleaned, or internal springs and shims renewed, using the approved, specialist equipment. However, if an injector is shown on the test rig to have excessive rate of leakage, the injector will have to be scrapped and renewed.

1 - illuminated spray chamber
2 - light switch
3 - high pressure pipe
4 - clock
5 - pressure gauge
6 - fuel control valve
7 - gauge isolator
8 - switch for extractor fan on/off
9 - filler cap
10 - fuel outlet
11 - hand lever
12 - sight tube
13 - tank drain
14 - inclined support
15 - quick action clamp

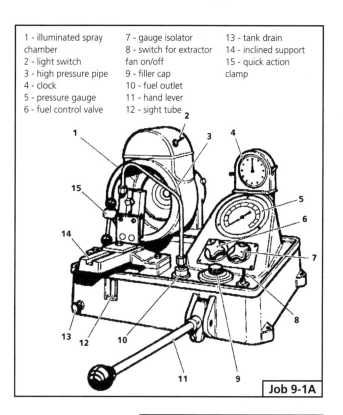

Job 9-1A

❏ **Step 1B.** Although the damage shown here is the result of poor spray pattern, the damage to this piston shows what can happen when injectors do not work properly.

Job 9-1B

FAULT 2: ENGINE MISFIRES, RUNS ERRATICALLY OR SURGES

Job 10. Restriction in induction system.

See *Job 7*.

Job 11. Wrong or incorrectly connected high pressure pipes.

❏ **Step 1.** Don't make assumptions! Check the manufacturer's manual for the correct pipe specifications, firing order and to see which cylinder is number one. On in-line engines the number one cylinder is usually nearest the crankshaft pulley but some manufacturer's differ, occasionally it's the one nearest the flywheel.

WORKING OUT THE CORRECT ORDER

❏ **Step 2.** On engines fitted with distributor pumps, make sure fuel is being delivered to number one cylinder at the correct time. Check the direction of pump rotation and make sure the pipes are connected in the correct firing order sequence.

• *During Step 2 the stop control will have to be in the ON position (i.e. engine can run), at least while number one cylinder approaches and passes TDC on its compression stroke. Although remote, there is a chance that the engine may fire, so it is a sensible precaution to disconnect all the injectors during this check. (However, note that marine engines are energised to STOP*

❑ **Step 3.** In-line injector pumps are easier to check since the outlet ports are usually connected to the respective cylinders, i.e. number one port goes to number one cylinder, number two port to number two cylinder and so on.

Job 12. Faulty lift pump.

Most lift pumps are of the diaphragm type but plunger types do exist. The comments here relate to the more common diaphragm type, the most usual problems with this type being a blocked internal strainer or punctured diaphragm.

❑ **Step 1.** Identify the pump type fitted to the engine. Some types, such as the one illustrated here have an internal filter mesh. At the manufacturer's specified intervals the top cover should be taken off, the filter mesh (**1**) removed, washed and refitted.

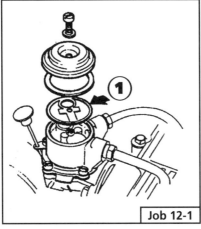

Job 12-1

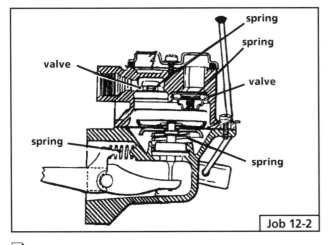

Job 12-2

❑ **Step 2.** The decision whether to repair or renew a lift pump will depend on the availability of repair kits. Generally, if the pump body is screwed together it can be dismantled and repaired. You should consult the manufacturer's manual, or any instructions supplied with the repair kit, for specific details on pump repair. Do watch out when dismantling pumps, they contain springs which can fly out during dismantling and valves which must be refitted correctly.

Job 14. Restricted fuel feed pipe.

Once you have found the offending section of pipe you will have to decide the best method of clearing the blockage. It's a good idea to completely remove the pipe otherwise attempts at clearing the blockage can just move it to another section of the system. Short sections of pipe can be cleared out with thin wire; longer sections are better cleared using compressed air. Alternatively, it may be easier to renew the blocked pipe. Make sure any replacement plastic or rubber pipes are suitable for use with diesel fuel. Whichever option is chosen, the fuel system will have to be bled - see *Job 2*.

Job 16. Incorrect valve clearances.

Before assuming incorrect valve clearances, do check to see whether your engine has hydraulic valve lifters. This type cannot be adjusted - although they can fail. On engines fitted with overhead camshafts, the cam usually operates the valves direct - there are no push rods and rockers. This type is more difficult to set but should stay in adjustment longer.

❑ **Step 1.** Consult the manufacturer's manual for the correct valve clearances. It is usually essential that they are checked with the engine cold - but in some cases, the engine has to be hot. The manual should also indicate the correct order for checking the clearances, which can vary between different engine types and engines with different firing orders.

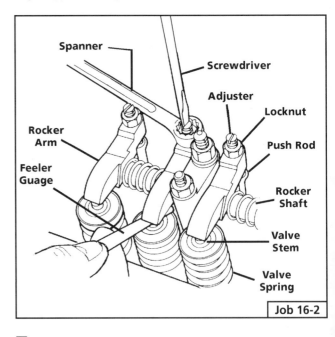

Job 16-2

❑ **Step 2.** Valves should be fully closed when the clearances are being set. On engines fitted with rockers, use a ring spanner to release the lock nut - when fitted.

Step 3. On engines fitted with overhead camshafts, washers or shims which sit directly on the valve stem are used to set the clearances. You have to measure the

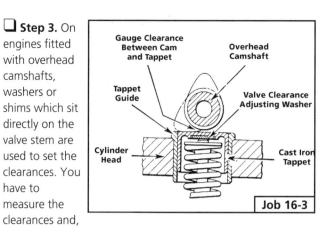

Gauge Clearance Between Cam and Tappet
Overhead Camshaft
Tappet Guide
Valve Clearance Adjusting Washer
Cylinder Head
Cast Iron Tappet

Job 16-3

clearances and, if any are incorrect, remove the camshaft. The thickness of the "incorrect" shim has to be measured and renewed with a shim of the correct thickness. A micrometer, or dial gauge and surface plate are needed to get a sufficiently accurate shim thickness measurement.

Step 4. Refit the camshaft and measure valve clearances - they should now be correct otherwise the process has to be repeated.

IMPORTANT NOTES: • If the engine has been rebuilt or the head gasket changed, the clearances should be checked before the engine is run.
• Traditionally, engines are re-checked after a short running time (specified by the manufacturer) but some modern engines - those which do not need the head to be re-torqued down or have overhead camshafts - do not need to be re-checked.
• All non-hydraulic types must be checked and, if necessary, re-set at the manufacturer's recommended service intervals.

Job 17.
Broken or weak valve springs.

Correcting broken or weak valve springs will require some dismantling. Fortunately, with most diesel engines it is possible to change valve springs without removing the cylinder head, although removing the rocker gear assembly is almost invariably necessary.

Step 1. Remove the rocker gear assembly. See **Chapter 3, Part C**.

Step 2. Rotate the engine until the piston below the chosen valves is at TDC. Using a suitable lever, compress the valve springs

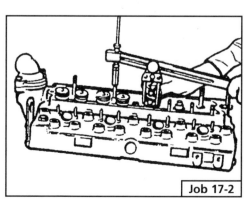

Job 17-2

(the valve will be held by the close proximity of the piston), then remove the collets and the springs. Keep a note of the position of any seals and spacers.

Step 3. Always renew the valve stem seals when changing valve springs.

Step 4. Fitting the new valve springs is the reverse of the removal process. Do make sure that any seals and spacers are put back in the correct order.

making it easy • *The valve collets are small and can easily fall down many of the openings in the cylinder head.*
• *Use rags or paper wipes to block all holes before compressing the valve springs.*
• *Grease can be used to hold the collets in place during reassembly.*

Step 5. Refit the rocker assembly and check the valve clearances.

Job 18. Sticking valves.

Step 1. Remove the cylinder head from the engine. See **Chapter 3, Part C**.

Step 2. Remove the valve springs (see **Job 17**) and identify the sticking valve or valves.

Step 3. Check the sticking valves to see whether they are bent or the problem has been caused by a build up of deposits. Excessive deposits in anything other than a very old engine indicates wear or poor adjustment.

Step 4. Rolling a valve on a flat surface may show up significant deformation. If you are not sure, try a new valve, or one that you know to be okay, in the guide of the suspect valve. Renew as necessary.

Step 5. Take the opportunity to check all valves and valve guides for wear. There is little point in cleaning or renewing valves if the guides are excessively worn. Use new valve springs if necessary (see **Check 17**) or if they are aged.

FACT FILE:
• Valve guides usually have to be pressed out and new ones pressed in.
• Valve guides often have to be reamed to the correct internal diameter (see manufacturer's specifications) once they have been pressed into the cylinder head.

Job 19.
Incorrect injection pump timing.

Unless the engine has recently been rebuilt or maintenance work, such as renewing the camshaft drive, has been undertaken it is unlikely that you will encounter this problem. However, after a very high mileage in-line injection pumps may require recalibration to make sure that the phasing is correct.

Because of the precision and critical assembly settings of the internal components, as well as the extent of special equipment needed for dismantling and reassembly, a thorough test, repair and calibration of a diesel injection pump should be entrusted to a trained expert with access to a dynamic pump test bench. While small, independent workshops may have the necessary experience and equipment to carry out the tests, pump manufacturers and suppliers encourage the use of their maintenance and support networks. The Perkins Powercentre network is a good example of this type of system.

A suspect pump returned to a Perkins Powercentre will be tested and, if necessary, repaired using approved components and calibrated to bring the settings back to those originally specified. Alternatively, the Perkins Powercentre can offer an exchange service which may be more cost-effective if engine down-time is critical.

There are some simple checks that can be carried out while the pump is still on the engine to determine whether the pump is suspect.

❏ **Step 1.** Check the fuel injection pump drive to make sure that the pump is being driven and the timing marks are correctly aligned (Static Timing).

❏ **Step 2.** Remove covers from over the timing belts, chains or gears.

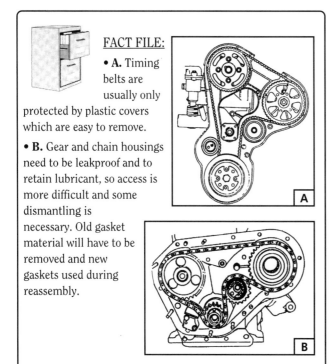

FACT FILE:
• **A.** Timing belts are usually only protected by plastic covers which are easy to remove.
• **B.** Gear and chain housings need to be leakproof and to retain lubricant, so access is more difficult and some dismantling is necessary. Old gasket material will have to be removed and new gaskets used during reassembly.

❏ **Step 3.** Turn the engine and set the timing marks as recommended by the manufacturer. Normally, the timing is set with No.1 piston at TDC on the compression stroke.

IMPORTANT NOTE: Remember, on a four stroke engine each piston compresses the air on every second stroke. It is a common mistake to set the timing marks, but with the engine at TDC on the wrong stroke. Any attempt to start the engine when set like this will almost certainly cause serious damage.

❏ **Step 4.** If the timing marks do not line up, follow the manufacturer's instruction for making the relevant adjustments.

❏ **Step 5.** If the engine runs for a short period it may be possible to check the dynamic timing. This check needs specialist equipment.

IMPORTANT NOTE: Incorrect injector opening pressures will affect injection timing, e.g. a low pressure will advance the timing. So, before jumping to the conclusion that the fuel pump is faulty, make sure the injectors are the right ones and are working properly. See *Job 9*.

Job 20. Incorrect valve timing.

The drive to the valves is normally shared with the drive to the injection pump, so the comments about the likelihood of incorrect injection pump timing also apply to incorrect valve timing. See *Job 19*.

IMPORTANT NOTE: If you make any adjustments to the valve timing, slowly rotate the engine by hand before attempting a restart. The clearance between pistons and valves is very small and what may appear as an insignificant misalignment of the timing marks (one tooth out) can be enough to bring them into contact. Turning the engine by hand is the least expensive way to confirm that pistons and valves are not going to collide.

Job 21. Incorrect or faulty injectors.

See *Job 9*.

Job 22. Engine does not run at correct operating temperature.

See *FAULT 7* and *FAULT 8*.

Job 23.
Poor cylinder compressions.

See *Chapter 3, PART C* for engine overhaul information.

Job 24. Speed control linkages stiff or sticking.

Speed control cables and linkages are often neglected parts of the engine and they can give problems which take time and effort to fix. Lubricate regularly, even if not part of the manufacturer's schedule (but DON'T use oil on nylon bushes - use silicon lube) and re-route cables with too sharp a bend, if possible.

PART A: CABLE TYPES

☐ **Step A1.** The solution may be quite simple, such as the linkage catching or rubbing, while a frayed or kinked cable inside a sheath is a more obscure problem.

• In the first case, re-routing the linkage may do the job.
• Alternatively, the cable may need oiling, which usually means disconnecting it and adding drips of oil while 'working' the cable sheath along the cable.
• If the cable is frayed or kinked, it should be renewed.
• Watch out for corroded cables and linkages.
• It may be possible to free off a corroded speed control but, unless the corrosion is cleaned away and the source of the trouble corrected, this problem will repeat itself. It's usually best to renew corroded cables.

PART B: ROD LINKAGE TYPES

☐ **Step B1.** Rod linkages can (and often do!) suffer from a number of faults, most of them caused by wear or corrosion:
• clevis (or other swivel) pins and bushes wear and accumulate lost movement
• trunnion mountings can come loose
• pins and bushes can be stiff and in need of lubrication, and sometimes, stripping and cleaning.

i INSIDE INFORMATION: With a rod linkage system, make sure that no lever is pulling too far over-centre in either direction. This will give a jerky response, sticking or incomplete movement. *i*

Job 27. Faulty governor.

A governor is an important part of the engine control system and, for safety, it is essential that the maximum speed setting is set correctly. Apart from setting the maximum and minimum speeds there is little else the user can do to confirm satisfactory operation. Visible springs and linkages can be checked for obvious signs of damage and seizure but, if the governor is considered faulty it should be removed from the engine for testing by a specialist workshop or approved service centre, such as a Perkins Powercentre. The following details are for general guidance only.

FACT FILE: GOVERNOR LOCATION

Most governors are integral with the fuel pump so a test, repair or recalibration of the pump will also include the governor.

However on some, such as the Perkins 100 Series, Perama and 700 Series engines, the governor is fitted to the camshaft. You should consult the appropriate Perkins manual for details on this governor.

The mechanical and hydraulic details and diagrams here relate to the CAV distributor-type injection pump which can be fitted with either a mechanical or hydraulic governor. The comments are for general guidance only and should not be used for servicing or maintenance - consult the manufacturer's manual.

MECHANICAL GOVERNOR

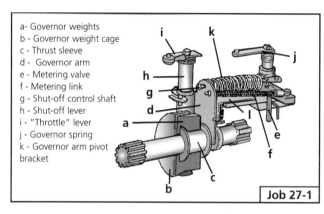

a- Governor weights
b - Governor weight cage
c - Thrust sleeve
d - Governor arm
e - Metering valve
f - Metering link
g - Shut-off control shaft
h - Shut-off lever
i - "Throttle" lever
j - Governor spring
k - Governor arm pivot bracket

Job 27-1

☐ **Point 1.** Maintaining a constant idling speed and limiting the maximum speed is taken care of by the governor. The principle of governing is based on the balance between the centrifugal force of the governor weights and the idling spring pressure. Differences in engine speeds brought about by changes in load are balanced out by adjusting the injection capacity. The governor induces a reduction in fuel supply when engine speed rises and an increase when engine speed falls. A mechanical governor has many components but this simplified diagram shows the most important.

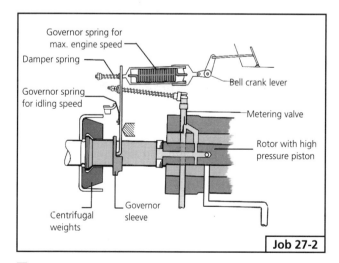

Governor spring for max. engine speed
Damper spring
Governor spring for idling speed
Bell crank lever
Metering valve
Rotor with high pressure piston
Governor sleeve
Centrifugal weights

Job 27-2

☐ **Point 2.** In a two-speed type of governor, where the governor only controls the minimum and maximum speeds, an additional pre-loaded main governor spring directly links the accelerator and metering valve at speeds above idle and up to maximum. This spring, sometimes called a 'mini-maxi' spring, works like a solid link until maximum speed is approached, when the centrifugal force of the governor weights overcomes the pre-load and the metering valve is turned to reduce the fuel delivery.

OPERATING PRINCIPLES

The governor weights (see illustration **Point 27-1, part a**) are contained in an open ended cage (**part b**). As engine and pump speed rise, increasing centrifugal force makes the weights move outwards. This movement forces the thrust sleeve (**part c**) and governor arm (**part d**) away from the weights and moves the fuel metering valve (**part e**) via the link (**part f**) to reduce the amount of fuel delivered.

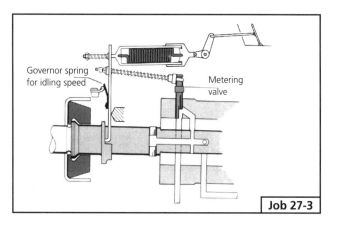

Governor spring for idling speed

Metering valve

Job 27-3

❑ **Point 3. ENGINE SPEED DROPS:** If, when the engine is idling, a power consumer is switched on, the idling speed drops briefly. The centrifugal weights change their position, the tension on the idling spring is now greater than the centrifugal force of the weights and the metering valve is opened. The engine speed increases until the centrifugal force of the weights and the idling spring tension are once again balanced out. The engine now runs at the set idling speed.

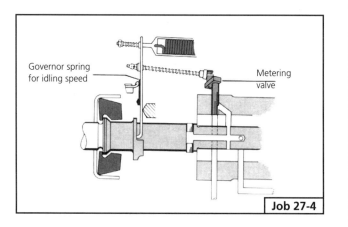

Governor spring for idling speed

Metering valve

Job 27-4

❑ **Point 4. ENGINE SPEED INCREASES:** If, when the engine is idling, a power consumer is switched off, or if the coefficient of engine friction is reduced, the engine speed increases briefly. This causes the centrifugal weights and the governor sleeve to change their positions, and the metering valve is closed slightly. The engine speed then drops. The idling spring relaxes and a balance is once again brought about. The metering valve allows sufficient fuel through to maintain the set idling speed.

❑ **Point 5. PART THROTTLE:** During part throttle operation, the engine speed governing is taken care of by means of the accelerator pedal only.

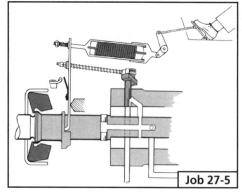

Job 27-5

❑ **Point 6. FULL THROTTLE:** To ensure that excessively high engine speeds do not lead to engine damage, the fuel supply is limited at a predetermined

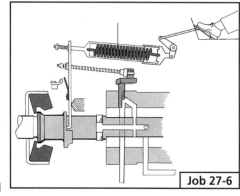

Job 27-6

top engine speed in the following manner. At high engine speeds, the centrifugal force of the governor weights overcomes the maximum engine speed spring tension and the metering valve is not opened any further.

HYDRAULIC GOVERNOR

Variations in engine and pump speed cause the fuel pressure in the pump to rise and fall. If this fuel, at "transfer" pressure, is channelled to a metering valve plunger, the pressure changes can be used to control the position of the metering valve, and the engine speed. However, hydraulic governors are not as sensitive as mechanical types and in these days of strict emission control are less common.

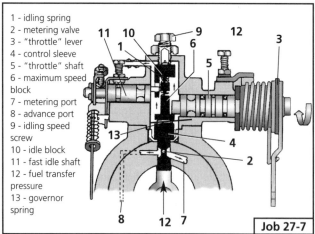

1 - idling spring
2 - metering valve
3 - "throttle" lever
4 - control sleeve
5 - "throttle" shaft
6 - maximum speed block
7 - metering port
8 - advance port
9 - idling speed screw
10 - idle block
11 - fast idle shaft
12 - fuel transfer pressure
13 - governor spring

Job 27-7

❑ **Point 7.** Although the governors described here are fitted to the same Lucas DPA pump, the first uses the shaft to act as a fast idle device...

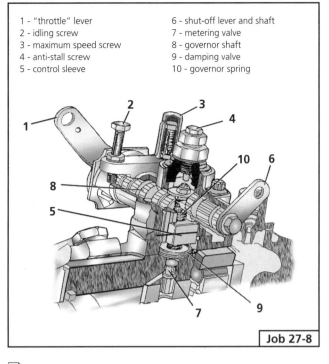

1 - "throttle" lever
2 - idling screw
3 - maximum speed screw
4 - anti-stall screw
5 - control sleeve
6 - shut-off lever and shaft
7 - metering valve
8 - governor shaft
9 - damping valve
10 - governor spring

Job 27-8

❑ **Point 8.** ...while the second uses this shaft (**part 6**) as a shut-off control. So beware! Similar looking devices may not behave in the same way. These differences are explained later.

When the engine is running, fuel at transfer pressure pushes the metering valve (see illustration **Point 27-7, part 2**) and control sleeve (**part 4**) upwards keeping the maximum speed block (**part, 6**) in contact with the eccentric pin on the end of the "throttle" shaft (**part 5**). This pin controls the position of the maximum speed block (**part 6**) (or just the control sleeve (see illustration **Point 27-8, part 5**)) above the metering valve plunger, in direct relation to speed control position. The exact position of the metering valve and control sleeve depend on the compression of the governor spring (**Point 27-7, part 13**), the fuel flow being determined by the amount by which the edge of the metering valve plunger uncovers the metering port (**part 7**) and the advance ports (**part 8**). If the "throttle" position is unchanged but engine speed increases, the metering valve rises against the governor spring which reduces the flow of fuel through the metering port and the engine speed falls. Any reduction in engine speed and the metering valve falls, increasing the flow of fuel.

Governed engine speed is changed by rotating the "throttle" lever, while movement of the eccentric pin changes the position of the maximum speed block, or the control sleeve, and the metering valve which rises or falls until fuel transfer pressure balances the compression of the governor spring.

ENGINE IDLING: At idle speed, the governor spring is fully extended and the eccentric pin is not in contact with the maximum speed block. Any movement of the metering valve is controlled by the idling spring (see illustration **Point 27-7, part 1**). Idle speed is adjusted by the screw (**part 9**), which alters the position of the idle block (**part 10**), or by the metering valve (**part 2**). Rotating the fast idle shaft (**part 11**) alters the position of the idle block (**part 10**).

ENGINE AT OPERATING SPEEDS: When the speed control is moved, the position of the control sleeve (see illustration **Point 27-7, part 4** or **Point 27-8, part 5**) and the metering

valve plunger are altered by the governor shaft eccentric pin. The position of the plunger edge moves to allow more or less fuel to flow into the metering drilling to the high pressure pumping section - and the engine speed changes until the "transfer" pressure underneath balances the spring pressure above.

At maximum speed the transfer pressure under the metering valve overcomes the downward force exerted by the governor spring plunger, while the metering valve rises, reducing fuel flow through the metering port and preventing further increases in engine speed.

IMPORTANT NOTE: You can see from this very brief overview of governor operation that there are many springs and linkages used to ensure correct mechanical and hydraulic governor operation. If there is a fault with one of these governors, it is very unlikely that you will be able to correct it in-situ. The fuel pump should be removed and sent to a diesel engine specialist, such as a member of the Perkins Powercentre network, for assessment.

FAULT 3: ENGINE VIBRATES EXCESSIVELY

Job 28. Sticking or restricted speed control movement.

See **Job 24**.

Job 29. Faulty engine mountings.

❑ **Step 1.** Steel beds and mounts may be repaired by welding or by fitting reinforcing plates, while damaged wood and glass fibre must be renewed or reinforced. To ensure that repairs have adequate strength, there are certain "rules" which should be followed but these are really beyond the scope of a book on engines. If you are not sure about carrying out structural repairs, you should seek guidance from specialists. Bear in mind that attempts to glue wood, or repair fibreglass with resin, are often thwarted by oily diesel fuel having seeped into the material and preventing adhesion.

SAFETY FIRST!

• *The temperatures reached during welding are more than sufficient to ignite diesel fuel, so adequate precautions should be taken to prevent fires starting in the first place, and to put them out if they do start. Seek specialist advice before welding in confined spaces and remove all flammable materials from the work area before carrying out welding.*

❏ **Step 2.** Broken flexible mounts and rubber mounts that have gone "soft" must be renewed. They can only rarely be repaired. Support or lift the engine so that the weight is off the mounting, remove the old mounting and fit the new one. Make sure that the new mounting is appropriate for the engine and that it is not stressed or loaded outside its design requirements.

Job 30. Damaged or loose fan or crankshaft pulley.

SAFETY FIRST!

• *In an emergency, it may be possible to continue using a damaged fan if blades are trimmed or selectively removed to balance any that are damaged or broken. Try to balance the fan statically while off the engine, then keep well clear in case of flying blades when the engine is running.*

Job 31. Wrong or incorrectly connected high pressure pipes.

See *Job 11*.

Job 32. Sticking valves.

See *Job 18*.

Job 34. Poor cylinder compressions.

See *Chapter 3, Part C* for engine rebuild information.

Job 35. Incorrect or faulty injectors.

See *Job 9*.

Job 36. Incorrect injection pump timing.

See *Job 19*.

Job 38. Leaking injector pipes (fuel escaping).

See *Job 3*.

Job 39. Incorrect valve clearances.

See *Job 16* and *Job 18*.

FAULT 4: ENGINE KNOCKS EXCESSIVELY

Job 41. Incorrect or faulty injectors.

See *Job 9*.

Job 42. Incorrect injection pump or valve timing.

See *Job 19* or *Job 20*.

Job 43. Incorrect valve clearances.

See *Job 16*.

Job 44. Broken valve springs.

See *Job 17*.

Job 45. Sticking valves.

See *Job 18*.

Job 46. Incorrect piston height.

This fault can only be detected and resolved by removing the cylinder head. See *Chapter 3, Part C*.

Job 47. Worn or damaged small-end or big-end bearings.

Step 1. This fault can only be confirmed by dismantling the engine and examining the condition of the journal and bearings or accurately measuring diameters and thicknesses. This damage is obvious. See *Chapter 3, Part C.*

Job 47-1

Job 48. Excessive camshaft bearing wear.

This fault can only be confirmed by dismantling the engine, inspecting camshaft and bearing visually and by accurately measuring them. See *Chapter 3, Part C.*

Job 49. Broken, worn or sticking piston rings

Step 1. This fault can only be confirmed by dismantling the engine and examining the condition of the piston rings and bores. Sometimes damage is obvious but, if no damage or wear is apparent, accurate measurement will confirm whether the components are serviceable. See *Chapter 3, Part C.*

Job 49-1

Job 50. Worn or damaged pistons.

Step 1. This fault can only be confirmed by dismantling the engine and examining the condition of the pistons. If no damage or wear is apparent, accurate measurement will confirm whether the components are serviceable. See *Chapter 3, Part C.*

Job 50-1

Job 51. Excessive timing gear backlash.

Depending on the design, and age, of an engine the camshaft and fuel pump may be driven by chains, flexible toothed belts or gears. Excessive backlash is most likely with chain drives - the chain stretches and wears beyond the capability of the tensioner to keep it tight. Toothed belts should be renewed at the manufacturer's recommended intervals which will be long before any wear will become critical - any noticeable backlash with a toothed belt drive will probably be due to incorrect adjustment. Gear drives may have noticeable backlash with the engine stationary and cold. However, as the engine warms and oil lubricates the gears this slight backlash may disappear. Always check the manufacturer's specification before assuming gear backlash is excessive.

A simple backlash check is to rotate the crankshaft backwards and forwards while watching the camshaft or fuel pump drive shaft. Anything more than a just perceptible delay means that backlash is excessive but see comments on gear drives.

Whichever type of drive you have, manufacturer's invariably recommend that the engine is rotated so that the timing marks on the pulleys, sprockets or gears are aligned before any dismantling takes place. There are often special tools available to lock the shafts in the correct positions but, with extreme care, these are not essential.

PART A: TIMING CHAINS

The first job is to find out what covers have to be removed to gain access to the timing chain. On some designs removing the rocker cover may be sufficient while on others the front cover may have to be removed. Since timing chains are lubricated by engine oil new gaskets will be required during reassembly. The next job is to find the join in the chain which may involve rotating the engine several times - the trick is to get the joining link in an accessible position with the timing marks aligned. Be careful when removing the joining link, there is usually a small spring plate which can fly off - usually into the sump.

If chain replacement is possible simply by removing the rocker cover, some fiddling with the old chain to get it out and with the new chain to make sure it routes correctly is often necessary, especially to make sure that the chain is tight on the drive side and all the slack is on the tensioner run. This has to be achieved while keeping the timing marks aligned.

Tensioning procedures will vary between manufacturer's. Some may specify that this is carried out with the engine stationary while others may specify a particular speed. Irrespective of the recommendation do turn the engine slowly by hand for two complete revolutions before starting - it is all too easy to slightly misalign the timing marks which can result in pistons hitting valves.

PART B: TIMING BELTS

Timing belts are usually easier to get at than chains since they are not lubricated and are usually only protected by plastic covers. However, because they are continuous, unlike chains, any ancillaries such as alternator drive belts, power steering drive belts etc. may also have to be removed.

The following procedure relates to the Perkins 500 Series and Prima engines and is shown for general information. It should not be used as a guide to other types of engine.

❑ **Step B1.** Disconnect the battery.

❑ **Step B2.** Remove the setscrew from the timing hole in the camshaft cover and turn the engine until the timing hole in the camshaft aligns with the hole in the cover. Fit the appropriate timing pins into the camshaft and flywheel.

❑ **Step B3.** Remove the starter motor and fit the anti-rotation tool to the flywheel

❑ **Step B4.** Remove the fan, alternator drive belt and water pump pulley.

❑ **Step B5.** Remove the access panel from the timing case, if fitted.

Job 51-B6

❑ **Step B6.** Depending on type, the timing cover may be in one, two, or even more parts.

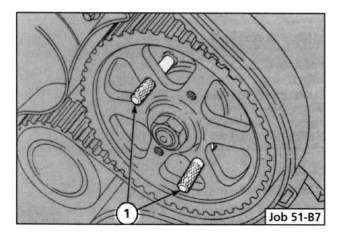

Job 51-B7

❑ **Step B7.** Fit the locating pins (**1**) through the plain holes in the pulley of the fuel injection pump and into the pump support bracket.

❑ **Step B8.** Slacken the belt tensioner pulley.

❑ **Step B9.** Remove the timing belt.

IMPORTANT NOTE: Toothed belts are very strong but can be damaged by mishandling. Make sure there are no cuts or nicks in the belt and do not bend the belt through very sharp angles - the resulting internal damage is not visible but can cause premature failure.

❑ **Step B10.** Fit a new belt, making sure that all the slack is on the tensioner run and any direction arrows point in the direction of rotation.

❑ **Step B11.** Remove the locking pins from the camshaft, flywheel and fuel pump pulley.

❑ **Step B12.** Most manufacturers recommend that a tension gauge is used when adjusting belt tension. This should be used at the longest run of the timing belt.

ℹ INSIDE INFORMATION: As a rule of thumb, most timing belts are correctly tensioned when it is just possible to twist the longest run of belt through 90 degrees using thumb and finger. Check with your engine's manual. ℹ

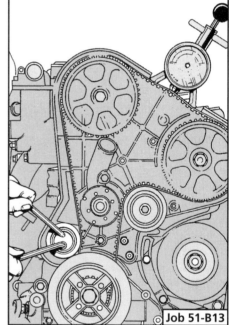

Job 51-B13

❑ **Step B13.** Use an Allen key (or the appropriate tool, depending on type) to adjust the belt tension to the correct figures and lock the tensioner.

❑ **Step B14.** Remove the tension gauge and turn the engine through two complete turns by hand to check for any obstructions.

❑ **Step B15.** Check the fuel pump timing.

❑ **Step B16.** Refit the covers and all other components removed to gain access.

PART C: TIMING GEARS

Unless the engine is very old, has been run with insufficient lubrication or has been incorrectly assembled during a rebuild it is very unlikely that the timing gear backlash will be excessive. It will normally be necessary to remove the timing cover to gain access to the timing gears. See **Chapter 3, Part C** for details on dismantling an engine with timing gears.

Step C1. Check the timing gears for wear or damage on the contact surfaces. Renew if any defect is visible.

Step C2. Use a dial gauge or thickness gauge to measure the backlash. For the Perkins 100 Series and Perama engines a standard backlash of 0.08 mm is quoted with an allowable limit of 0.25 mm. If the backlash exceeds the limit the gears should be renewed. See *Chapter 3, Part C*.

Job 52. Overfuelling through faulty or sticking governor or speed control system.

See *Job 24* and *Job 27*.

Job 53. Faulty lift pump.

See *Job 12*.

Job 54. Faulty cold starting equipment.

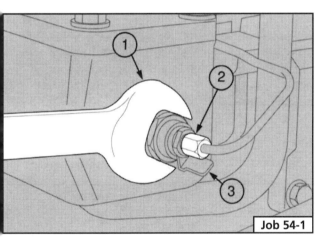

Job 54-1

Step 1. If a glow plug fails, it can only be renewed, not repaired. However, the electric (**3**) and fuel (**2**) connections to the Perkins 'Thermostart' system can be checked and the unit can be removed (**1**) for physical examination.

Job 55. Overheating.

See *FAULT 8*.

Job 56. Piston seizure/pickup.

This fault can only be confirmed by dismantling the engine and examining the condition of the pistons and cylinder bores. If no damage or wear is apparent, accurate measurement will confirm whether the components are serviceable. See *Chapter 3, Part C*.

FAULT 5: ENGINE WILL NOT ACCELERATE OR PRODUCE POWER

Job 57. Stop control not in fully ON position.

See *Check 25*.

Job 58. Engine speed control sticking or movement restricted.

See *Job 24*.

Job 59. Restriction in induction system.

See *Job 7*.

Job 60. Air in fuel system.

See *Job 2*.

Job 65. Incorrect injection pump or valve timing.

See *Job 19* or *Job 20*.

Job 66. Restriction in exhaust system.

See *Check 8*.

Job 67. Faulty, worn or incorrect injectors.

See *Job 9*.

Job 68. Faulty, worn or incorrect injection pump.

There is very little that the "user" can do if an injection pump is suspected of being faulty or worn. An incorrect pump should be renewed.

Specialist equipment, such as dynamic test bench, is needed to properly check and calibrate an injection pump. The necessary equipment should be available in a specialist diesel engine repair shop but to be sure the work can be entrusted to a member of the engine manufacturer's service network, such as the Perkins Powercentre organisation.

See *Job 19*.

Job 69. Incorrect valve clearances.

See *Job 16*.

Job 70. Poor cylinder compressions.

See *Chapter 3, Part C*.

Job 71. Poor boost pressure (turbocharged engines)

Thorough turbocharger testing requires specialist equipment. However, there are some tests that can be carried out fairly easily. Beware though turbochargers get very hot so any physical examination of the turbocharger should be done with the engine cold.

❏ **Step 1.** With the engine **COLD** remove the air inlet to the turbocharger and examine the compressor vanes. Look for any damage, erosion or contamination. A damaged or worn compressor and the turbocharger will have to be renewed. Contamination can be cleaned but check the manufacturer's recommendations for acceptable cleaning materials and methods.

❏ **Step 2.** Rotate the compressor, listening and feeling for any binding. Check for any free play. If anything is found the turbocharger will have to be renewed.

❏ **Step 3.** A visual check of the wastegate valve is more difficult because the exhaust pipe will have to be disconnected at its inlet to the turbocharger. Depending on the particular design it may, or may not, be possible to get a good view of the valve and, in any case, it will probably be difficult to spot a valve that is only stuck open a small amount. Use an air pressure gauge connected to the inlet manifold between the turbocharger and the engine to check the boost pressure and operation of the wastegate. You will need to consult the

manufacturer's manual for details of acceptable pressures and wastegate operating conditions.

Any further testing of the turbocharger will require removal and return to a specialist such as a member of the Perkins Powercentre network.

Job 72. Sticking fuel delivery valves.

Fuel delivery valves must only be checked or renewed by the supplier or support centre. Delivery valves are fitted to the injection pump outlets and can be easily removed.

❏ **Step 1.** Remove the delivery valves and check that they are the ones specified by the manufacturer.

❏ **Step 2.** Refit overhauled or new delivery valves and reconnect the high pressure delivery pipes.

FAULT 6: EXCESSIVE LUBRICATING OIL CONSUMPTION

Job 73. Oil leak.

Some leaks are easy to fix while others involve major dismantling. The unions to separate oil coolers can be tightened and this usually solves the problem. Leaking pipes should be renewed.

Tightening attachment bolts can sometimes cure oil leaks between components but generally only if they were loose to start with and if gaskets are undamaged. Overtightening, in an attempt to stop a leak, doesn't often work and can damage or distort components. Reassembly with new gaskets or jointing compound is the only permanent cure.

External pipes work-harden over a lengthy period of time, because of engine vibration and can split, especially near their ends. Renew if necessary.

Job 75. Engine breathing system blocked.

Although most engines have crankcase breathing systems, the designs vary from engine to engine.

❏ **Step 1.** Check the engine or manufacturer's manual to find out the type of system fitted to your engine, the routing of pipes and the location of any valves and filters.

❏ **Step 2.** Some systems have scavenge pumps fitted, which assist the flow of combustion products from the crankcase. These may be vacuum or pressure operated pumps. If the pump is not working correctly, check to see if it can be restored by cleaning it, and renew if necessary.

Step 3. If excess fumes and a resultant blockage are caused by bore wear, a general overhaul will be required. It has been known for a diesel engine to fume so badly that it runs on its own lubricating oil and revs. itself to destruction. Overhaul the engine before it comes to this! See *Chapter 3, Part C*.

Job 76. Worn valve stems or valve guide bores.

Step 1. Worn valve stems or valve guide bores can only be confirmed by dismantling. At the very least the valve springs will have to be removed, which can be done with the cylinder head in place. If there is any perceptible side to side movement once the springs have been removed, the wear is excessive.

Step 2. Remove the cylinder head and remove the valves. Use a micrometer to measure the valve stem diameters and compare the figures with the manufacturer' specifications. If any wear is evident, they will have to be renewed.

Step 3. See *Job 18* for details of replacing valve guides.

Job 77. Worn valve stem seals.

Step 1. Check the manufacturer's manual to see if any special tools are needed. Remove the rocker cover and rocker assembly.

SAFETY FIRST!

• *Wear suitable eye protection when removing valve collets!*

Step 2. Decide which cylinder to work on and turn the engine until the piston is at TDC. Use a lever type compressor to compress the springs. Remove the collets, springs and old seals. Keep a note of any seats or spacers.

making it easy   • *The valve collets are small and can easily fall down many of the openings in the cylinder head.*
• *Use rags or paper wipes to block these holes before compressing the valve springs.*
• *Grease can be used to hold the collets in place during reassembly.*

Step 3. Check the valves to see if there is any noticeable valve stem or valve guide wear. If so, the cylinder head will have to come off for overhaul. See *Job 18* and *Job 76*.

Step 4. Fit new valve seals and, preferably, new springs.

Step 5. Repeat the process for each cylinder, reassemble the rocker gear, check the valve clearances and fit the rocker cover.

Job 78. Worn cylinder bores or pistons.

See *Chapter 3, Part C*.

Job 79. Worn or broken piston rings.

See *Chapter 3, Part C*.

Job 82. Glazed cylinder bores.

This fault can only be confirmed by dismantling the engine and examining the condition of the cylinder bores. Glazed cylinder bores must be honed to achieve a slightly rough finish.

Step 1. Dismantle the engine. See *Chapter 3, Part C*.

IMPORTANT NOTE: Special equipment and a skilled operator will be necessary when honing the bores of a modern engines because of the exacting requirements of modern emission laws.

Step 2. Before honing the bores, check their diameters with an internal micrometer. There is little point in honing them only to find that a rebore or reline is needed. There are several types of honing machine available and you should follow manufacturer's instruction when using them. Don't hone excessively, as a small amount of metal is removed each time the machine passes up and down the bore and you might remove too much material.

IMPORTANT NOTE: Honing is an abrasive process and produces highly abrasive debris. Make sure the engine and particularly the oilways are thoroughly cleaned out after this activity.

Job 85. Oil leaks from ancillary equipment.

Some dismantling and replacement of oils seals or gaskets will be needed to fix this problem, following the relevant manufacturer's manual.

FAULT 7: ENGINE RUNS COLD

Job 87. No thermostat, faulty thermostat, or wrong thermostat fitted.

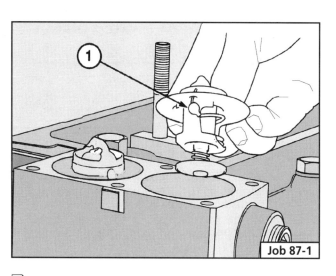

Job 87-1

❑ **Step 1.** Remove the thermostat housing and check to see whether a thermostat (**1**) is fitted. If one is fitted, check to make sure it is the correct type and specification and that it is working properly. See **Check 87** for thermostat testing.

❑ **Step 2.** Fit a new thermostat or refit the existing thermostat if it seems serviceable.

Job 88. Faulty gauge or transmitter.

❑ **Step 1.** Component repair is not normally possible and renewal is the best option. Follow the tests detailed in **Check 88** and you will have a good idea of which components are faulty.

❑ **Step 2.** If you have to follow a policy of test-by-replacement, renew the transmitter first (it's usually cheaper) and see if this make any difference. If not, you can fit a replacement gauge or try dismantling it to adjust the pointer.

FAULT 8: ENGINE OVERHEATS

Job 89. Insufficient coolant.

SAFETY FIRST!

• *Do not attempt to remove the filler cap or dismantle any part of the cooling system when the engine is hot. Engine cooling systems are normally pressurised which prevents the coolant boiling until a higher temperature than if it was at atmospheric pressure. Depressurising the system can allow the coolant to boil, with steam and boiling liquid spraying out of the opening at considerable force.*

IMPORTANT NOTE: Do not top up a hot engine with cold water - the thermal shock can sometimes be sufficient to cause internal cracks.

After a period of time, hose clips will rust solid and hoses will vulcanise themselves to stubs. Carefully cut through each clip with a junior hacksaw, then cut through the hose, at the stub, so you can peel it away.

making it easy • *Lubricate the knife blade with soapy water. It will then cut through rubber far more easily.*

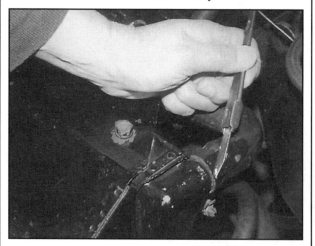

• *Use some soapy water or silicone lubricant on the pipe stubs when fitting new hoses.*
• *Getting all the air out of the cooling system when it is being refilled can be difficult and may take some time. Run the engine and check the water level frequently until it stabilises.*

Job 90. Oil level too high.

See **Check 74**.

Job 91. Faulty thermostat.

See *Job 87*.

Job 92. Blocked coolant system, faulty or incorrect radiator, coolant hoses or pressure cap.

See *Job 89* for useful tips when fitting replacement hoses.

MARINE ENGINES

Check that the water inlet is not fouled. Also, dismantle and check the heat exchanger (if fitted) for fouling.

SALT WATER HEAT EXCHANGERS: Engine coolant passes through the tube pack (**a**) while sea-water is pumped around the tubes. The two are separated by the sealing rings (**b**). The tubes can be cleaned without further dismantling by removing the end cap (**c**). However, tubes can become fouled with hard deposits, on either the coolant or sea-water sides and in this case, you will have to dismantle the whole heat-exchanger.

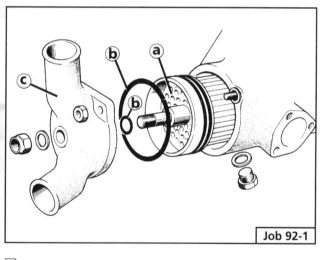

Job 92-1

❑ **Step 1.** Fit new seals on reassembly.

Job 94. Faulty water pump.

❑ **Step 1.** Drain the coolant so that the level is below the water pump and slacken the fan belt. Often the water pump is on the same shaft as the fan, so this may have to be removed to gain access and, depending on engine design, the location of other engine ancillaries may make the job of water pump removal less than easy.

❑ **Step 2.** Unbolt and remove the water pump.

❑ **Step 3.** Clean off any old gasket material from the mating face on the engine and fit a replacement water pump. Always use a new gasket.

❑ **Step 4.** See if there are any part numbers stamped or cast onto the replacement pump body. These should be compared with the manufacturer's specification.

❑ **Step 5.** ℹ INSIDE INFORMATION: If there is too small a pulley on the pump, it will rotate faster than necessary, wasting energy and possibly causing cavitation which reduces pump efficiency and increases wear. A pulley that is too large will cause the pump to run slower than design, which may also cause overheating, particularly at low speeds. Check with the manufacturer to see what size of pulley is recommended - although it is the ratio between the drive and driven pulleys which is critical. ℹ

Job 95. Leaking cylinder head gasket or cracked cylinder head.

Having carried out the checks in *Check 95*, three further engineering shop checks will have to be carried out in order to determine the appropriate action.
• test the cylinder head face for trueness
• test the block top face for trueness
• pressure test the head for cracks or porosity.

Cylinder head or block refacing is normally a straightforward job.

ℹ INSIDE INFORMATION: On some engines, you will have to use an appropriate gasket thickness to re-establish the correct compression ratio and/or piston projection. A cracked or porous cylinder can usually be (specialist) repaired, but a replacement is usually less expensive. ℹ

Job 96. Incorrect injection pump or valve timing.

See *Job 19* or *Job 20*.

Job 97. Faulty or incorrect injectors.

See *Job 9*.

Job 98. Restriction in induction system.

See *Job 7*.

Job 101. Faulty gauge or temperature transmitter.

See *Job 88*.

FAULT 9: OIL PRESSURE TOO LOW

Job 103. Worn or damaged bearings.

See *Job 47*.

Job 106. Worn oil pump.

Dismantling and accurate measurement of the pump's internal clearance are necessary before this fault can be confirmed. See *Chapter 3, Part C* for pump removal and consult the manufacturer's manual for guidance.

Job 107. Pressure relief valve stuck open.

If you are lucky you may be able to gain access to the relief valve without dismantling the engine. Consult the manufacturer's manual to see where the valve is located.

❏ **Step 1.** Remove the relief valve and check to see whether it has stuck. On some engines dismantling is the only way of removal and on most occasions the very act of removing the valve will cause it to free off.

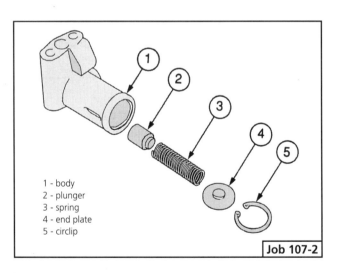

1 - body
2 - plunger
3 - spring
4 - end plate
5 - circlip

Job 107-2

❏ **Step 2.** Examine the dismantled components for any signs of uneven wear or sticking. Sometimes, debris in the sump can cause the valve to stick so check the sump oil for any serious contamination.

❏ **Step 3.** If nothing is found, reassemble and refit the valve. Otherwise, any stuck or damaged components should be renewed.

Job 108. Broken relief valve spring.

See *Job 107*.

Job 109. Faulty suction pipe.

Very few problems are encountered with suction pipes but they can be blocked by gross contamination or they can work loose. Dismantling is necessary; at least the sump will have to be removed, before this fault can be confirmed.

Job 110. Blocked oil filter.

Most oil filters have by-pass valves so that if the filter is blocked unfiltered oil can get to the bearings. There is no real way of checking whether a filter is blocked and renewal is the only practical solution - filters are cheaper than engine overhauls.

Some exotic types of filter may have a small indicator button which pops out if the filter becomes choked but this type is rare. Others may have an electric switch which operates a warning light if the filter becomes blocked. All types of oil filter have by-pass valves which will allow unfiltered oil to go to the engine if the filter becomes blocked but this will cause premature wear and is an emergency failure measure. Always renew the filter at the recommended intervals.

❏ **Step 1.** Remove the old oil filter and be ready for some spillage so have a drip tray or container ready. Replacement element filters can be removed easily with a spanner but usually a strap or chain wrench is needed to remove the throwaway type. Make a note of any spacers, springs or washers that may be used with replacement element filters.

❏ **Step 2.** Clean off the filter seat and, if necessary clean out the container for the replacement element type.

❏ **Step 3.** Fit a new oil filter canister or renew the element before reassembling the filter container. Smear a small quantity of oil on the sealing washer to prevent it sticking and distorting. Use new seals, if separate seals are used.

❏ **Step 4.** Do not overtighten. Hand tight plus 1/4 to 1/2 turn should be adequate.

❏ **Step 5.** Run the engine for a few minutes and check for leaks.

❏ **Step 6.** Stop the engine and allow it to stand for a few minute. Check the oil level and top up as necessary.

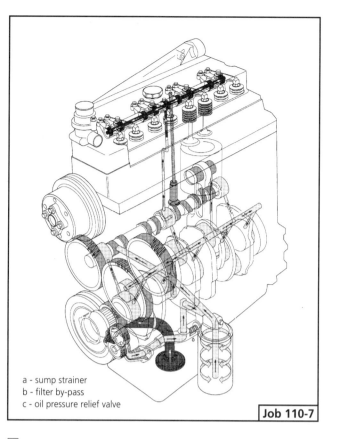

a - sump strainer
b - filter by-pass
c - oil pressure relief valve

Job 110-7

☐ **Step 7.** Renew the oil and filter at the periods specified by the manufacturer or investigate the reason for the blockage if the problem occurs within the operating period specified by the manufacturer. As this drawing shows, the smaller oil passages are the ones most at risk of blockage.

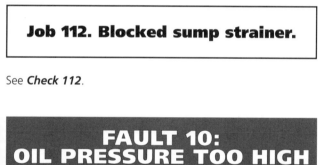

Job 112. Blocked sump strainer.

See *Check 112*.

FAULT 10: OIL PRESSURE TOO HIGH

Job 114. Faulty gauge or pressure transmitter.

See *Check 105*.

Job 115. Pressure relief valve sticking closed.

See *Job 107*.

FAULT 11: ENGINE OVERSPEEDS

Job 116. Incorrect maximum speed setting.

SAFETY FIRST!

• *Do not attempt to check or reset the maximum speed by guesswork - it cannot be done. An over-revved engine may disintegrate with disastrous consequences for the engine and nearby personnel. Always use a reliable tachometer when checking maximum engine speed.*

☐ **Step 1.** Identify the type of governor fitted to your engine and follow any instructions given in the manufacturer's manual. Usually mechanically and hydraulically governed fuel pumps have simple maximum speed screw adjustments, but do make sure you adjust the correct screws.

☐ **Step 2.** Setting pneumatic governors is a little more complicated and a general procedure is as follows: Start the engine and unscrew the damping valve until the engine surges. Gradually open the venturi butterfly until it is fully open. Note the engine speed and, if this is not correct, re-adjust the maximum speed stop on butterfly. Screw in the damping valve until the engine just stops surging and lock the damping valve.

Job 117. Faulty or incorrectly set governor.

See *Job 27*.

Job 118. Faulty or incorrectly set injection pump.

See *Job 68*.

FAULT 12: ENGINE WILL NOT STOP

Job 119. Faulty, damaged or incorrectly set stop mechanism.

See *Check 25*.

Job 120. Fuel or oil leak into induction system or cylinders.

See *Job 3* or *Job 73*.

FAULT 13: EXCESSIVE FUEL CONSUMPTION

Job 121. Fuel leaks.

See *Job 3*.

Job 122. Restriction in induction system.

See *Job 7*.

Job 123. Sticking valves.

See *Job 18*.

Job 124. Incorrect valve clearances.

See *Job 16*.

Job 125. Faulty cold starting aid.

See *Job 54*.

Job 126. Incorrect injection pump or valve timing.

See *Job 19* or *Job 20*.

Job 127. Faulty or incorrect injectors.

See *Job 9*.

Job 128. Faulty or incorrect injection pump.

See *Job 68*.

Job 129. Poor cylinder compressions.

See *Chapter 3, Part C* for engine rebuild information.

CHAPTER 5
SMOKE & EMISSION

As legislation becomes ever stricter and as operators become more aware of the fact that excessive exhaust emissions are harmful, there is a greater emphasis on minimising the harm done by exhaust emissions. And non-highway engines are affected just as much as their automotive brethren!

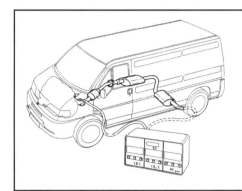

There are two main reasons why a diesel engine will produce smoke or excessive emissions: either vital components are very badly worn, or they have been set incorrectly. But don't make hasty assumptions! Use a methodical and systematic approach which will, in the longer term, save time and money and ensure you have better success at finding that fault.

How To Use This Chapter

This Chapter is divided into three parts:
• PART A lists possible faults and remedies.
• PART B shows you how to check for each fault.
• PART C deals with how to carry out necessary repairs.

Each fault has the same 'Check' number in PART A and PART B, and the same number (now a 'Job' number) in PART C.

Chapter Contents

Smoke is caused either by excess fuel for the available air or by oil burning in the combustion chambers. Excessive emissions are caused by poor combustion of the fuel. The two are related and, it may be argued, the white start-up smoke produced by some engines when they are cold could be classed as an emission problem rather than smoke. Many of the faults already listed will contribute to smoke or excessive emissions, so although this Fault Finding Checklist may seem familiar, it has been compiled with the diagnosis and repair of smoke and emission problems specifically in mind.

ENVIRONMENT FIRST!

• *The public is becoming increasingly aware of the contribution that internal combustion engines, including diesel engines, make to air pollution.*

• *Governments in many parts of the world have responded to these concerns and have introduced strict limits and checks on the acceptable levels of smoke and emissions from diesel engined road vehicles. Limits on non-automotive diesel engines are also being developed and enforced.*

• *There is legislation in many countries relating to smoke from industrial plant and the relevant authorities in your territory should be contacted for further advice.*

FACT FILE: TYPES OF SMOKE

There are, broadly speaking, three types of visible excessive smoke emissions which can give a guide to engine malfunction. Of course, if there is more than one type of fault, there will be more than one type of visible smoke - but you have to start somewhere.

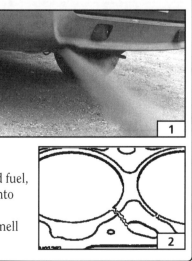

• **1.** BLACK (OR DARK GREY) SMOKE is a symptom of incompletely burnt fuel passing out of the engine and can be caused by low combustion pressures and temperatures (which is why diesel engines smoke on start-up) as well as by the faults shown below. Some of the fuel burns and the remainder decomposes to form sooty particles.

• **2.** WHITE, OR GREYISH-WHITE SMOKE is caused either by completely unburned fuel, in the form of vapour, or vaporised lubrication oil. It is possible for coolant to leak into the combustion chambers, via a cracked cylinder head or leaking head gasket, and produce steam which may be confused with white smoke. Smoke persists and will smell "oily"; steam disappears quickly and has little smell.

FACT FILE: TYPES OF SMOKE continued

• **3.** BLUE, OR BLUE-GREY smoke is caused by lubricating oil being burned in the combustion process - as the result of an engine defect, excess engine oil or incorrect type of engine oil in use.

• **4.** If the engine has to pass a statutory test, visible smoke limits may apply to some older engines (depending on the context in which it is used) but newer engines may be subject to a smoke meter capable of accurately measuring the volume of particulate emissions.

• **5.** This is a smoke opacimeter in use.

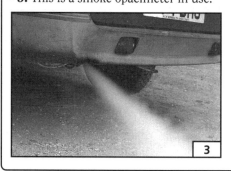

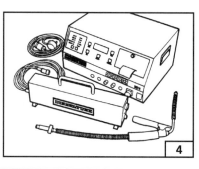

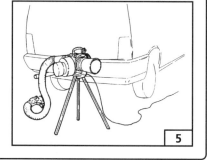

PART A: FAULT FINDING CHECKLISTS

FAULT 1: ENGINE PRODUCES BLACK (OR DARK GREY) SMOKE

	SYMPTOM	LIKELY CAUSE	REMEDY
Check 1.	❏ Black or grey smoke under load, more noticeable at high speed.	❏ Induction system restriction.	❏ Remove restriction.
Check 2.	❏ Black or grey smoke under load, more noticeable at high and medium speeds. Engine may lack power.	❏ Dirty, worn or incorrect injectors.	❏ Clean or replace injectors
Check 3.	❏ Black or grey smoke under load, more noticeable at high and medium speeds. Engine runs quieter than normal.	❏ Retarded fuel injection pump timing.	❏ Reset timing. Overhaul or replace fuel injection pump if condition is suspect.
Check 4.	❏ Black or grey smoke under load, more noticeable at low and medium speeds. Engine is noisier than normal.	❏ Advanced fuel injection pump timing.	❏ Reset timing. Overhaul or replace fuel injection pump if condition is suspect.
Check 5.	❏ Black or grey smoke at all speeds, more noticeable at low and medium speeds. Starting difficult.	❏ Poor cylinder compressions.	❏ Overhaul engine.
Check 6.	❏ Intermittent black or grey smoke. Maybe associated with a misfire or knock.	❏ Sticking injector.	❏ Repair or replace injector.
Check 7.	❏ Black or grey smoke at all speeds. Noticeable lack of power.	❏ Poor boost pressure (turbocharged engines).	❏ Overhaul boost control system.
Check 8.	❏ Black or grey smoke, most noticeable at low and medium speeds. Engine may also misfire.	❏ Incorrect valve clearances.	❏ Adjust valve clearances.

Check 9. ❑ Black or grey smoke at all speeds. Engine may also misfire or knock	❑ Incorrect type or grade of fuel.	❑ Drain system and refill with correct fuel.
Check 10. ❑ Black or grey smoke at all speeds, with engine warm. Most noticeable at low and medium speeds.	❑ Faulty cold starting equipment. (Engine fuel types)	❑ Repair or replace cold start equipment.

FAULT 2: ENGINE PRODUCES WHITE OR GREYISH WHITE SMOKE

SYMPTOM	LIKELY CAUSE	REMEDY
Check 11. ❑ Whitish smoke at high engine speeds and light loads. Possible unusual exhaust smell.	❑ Cold running.	❑ Fit or replace faulty thermostat.
Check 12. ❑ Whitish smoke at high engine speeds and light loads. Most noticeable when engine is cold. Colour changes to black as engine temperature rises.	❑ Retarded fuel injector pump timing.	❑ Reset timing. Overhaul or replace fuel injection pump if condition is suspect.
Check 13. ❑ Whitish smoke at light loads and engine at normal operating temperature. May be accompanied by knocking.	❑ Leaking injector or injectors.	❑ Repair or replace suspect injectors.

FAULT 3: ENGINE PRODUCES BLUE OR GREYISH BLUE SMOKE

SYMPTOM	LIKELY CAUSE	REMEDY
Check 14. ❑ Blue smoke on acceleration after the engine has been idling.	❑ Leaking valve seals or worn valve guides and valve stems.	❑ Replace seals or overhaul cylinder head.
Check 15. ❑ Blue smoke at all speeds, loads and temperatures.	❑ Worn piston rings or cylinder bores.	❑ Overhaul engine.
Check 16. ❑ Blue smoke at high speeds, low to medium speeds and all loads and temperatures.	❑ Leaking turbocharger oils seals.	❑ Overhaul turbocharger.
Check 17. ❑ Blue smoke at high temperatures and all speeds and loads.	❑ Lubricating oil too thin.	❑ Drain system and refill with correct grade of oil.

FAULT 4: EXCESSIVE SMOKE OR VAPOUR FROM CRANKCASE VENTILATION OUTLET

SYMPTOM	LIKELY CAUSE	REMEDY
Check 18. ❑ Blue or grey blue smoke from crankcase ventilation outlet.	❑ Worn piston rings or cylinder bores.	❑ Overhaul engine.
Check 19. ❑ Excessive oil vapour or oil droplets from crankcase ventilation outlet.	❑ Defective exhauster (if fitted)	❑ Overhaul vacuum system.

PART B: FAULT FINDING STEP-BY-STEP

FAULT 1: ENGINE PRODUCES BLACK (OR DARK GREY) SMOKE

Check 1.
Induction system restriction.

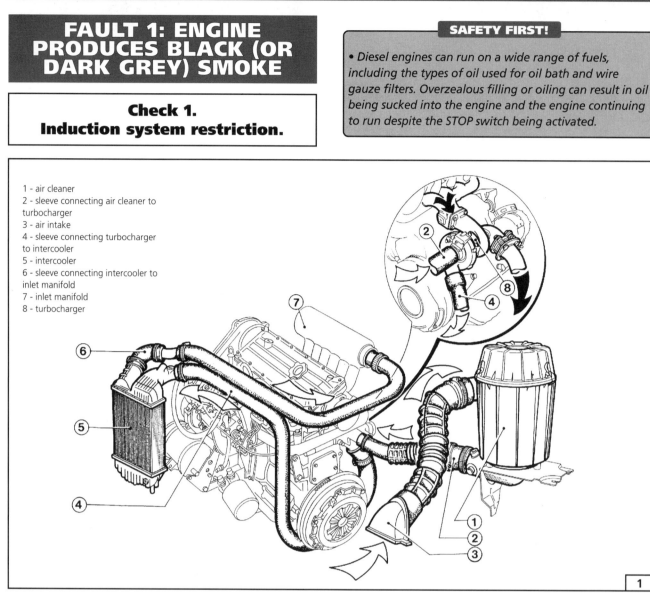

1 - air cleaner
2 - sleeve connecting air cleaner to turbocharger
3 - air intake
4 - sleeve connecting turbocharger to intercooler
5 - intercooler
6 - sleeve connecting intercooler to inlet manifold
7 - inlet manifold
8 - turbocharger

1. A restriction in the induction system will reduce the amount of air sucked into the cylinders during each induction stroke, which will reduce the power output of the engine. Trying to maintain power or speed, the operator will 'open the throttle' which increases the amount of fuel injected into the combustion chambers. More fuel and less air means that some of the fuel is simply turned into sooty carbon particles which come out of the exhaust as black smoke. These are the components of a typical (Fiat) light commercial vehicle's air intake circuit on an engine with turbocharger.

Apart from the environmental aspects and the waste of fuel, the particles will deposit themselves inside the engine. They will find their way into the oil, reducing its effective life and they will gradually build up on internal engine surfaces, causing hot spots, sticking valves and sticking piston rings. An engine that produces a lot of black smoke will cost more to run and need overhauls more frequently than one that is kept properly adjusted and in good condition.

A restriction in the induction system is often simply due to a blocked air filter, particularly the paper element type. Any build up of dirt, or the element becoming wet, will restrict the air flow.

FACT FILE: EXHAUST LEAKS

Watch out for the danger of leaking exhausts, especially if the engine is run in a confined space such as in a boat.

• If the exhaust leaks, some exhaust gas can be sucked back into the engine, contaminating the air filter and increasing the smoke and emissions from the exhaust.

• A vicious circle is set up, the more the filter becomes contaminated and blocked the worse emissions become, blocking the filter even more.

Blocked paper element filters normally have to be replaced but other types, such as fabric element, oil bath or wire gauze filters can be cleaned. The manufacturer's manual should provide guidance on cleaning procedures. If you have any doubts about the condition of the filter, measure the air pressure in the manifold or, more easily, disconnect the filter to see if this makes any difference. Do not run the engine without an air filter for any length of time - it's surprising how much premature bore and piston ring wear will take place.

Some industrial engines have an air flow shut-off valve. Check to see if there is one fitted and make sure it is in the fully open position. If an emergency air shut off valve has been activated, find out why before attempting to start the engine.

If a new or cleaned filter is fitted and the problem persists, check the intake ducting for both damage and blockages. Convoluted intake pipes can collapse internally - and it's surprising where mice and birds will build their nests when an engine has not been run for any length of time!

Change the air filter at the manufacturer's recommended intervals, or more frequently if the operating conditions are very dusty, such as around construction work. If a paper element filter is found to be wet, investigate the reason and take action to prevent it from happening again.

In an emergency, run the engine without a filter but be aware of the damage and rapid wear that unfiltered intake air will cause.

Check 2.
Dirty, worn or incorrect injectors.

2. For correct and efficient combustion it is vital that injectors spray finely atomised fuel into the combustion chamber at the right time. A dirty, worn or incorrect injector can cause large drops of fuel to spray or dribble into the combustion chamber, often at the wrong time.

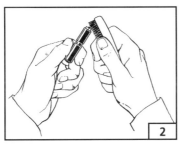

Check the manufacturer's specifications for the correct type of injector. Use an enclosed tester to check the function of each injector. Replace any that do not match the specification and any that do not function correctly. Make sure that the correct sealing and insulating washers are fitted.

IMPORTANT NOTE: To get the longest possible life from injectors, make sure that the fuel entering the injection pump is properly filtered. If the high pressure pipes are disconnected at any time, ensure that no dirt can get into the system and make sure that filters are in good condition.

SAFETY FIRST!

• *Diesel fuel is injected at very high pressures, so wear appropriate eye protection and do not let the fuel spurt or spray onto any part of the body.*
• *The pressures are high enough to force the fuel through the skin into the bloodstream with possibly fatal consequences.*
• *In particular, some modern common rail injection systems work at extremely high pressures - 1,350 bar in the Fiat and Mercedes Benz systems.*
• *Leave very high-pressure systems to stand, without the engine running, for the period of time recommended by the manufacturer. This allows the extremely high pressure to be reduced, before attempting to depressurise the system.*

Check 3.
Retarded injection pump timing.

If the injection pump timing is retarded, fuel will be injected after the optimum point and there will not be enough time for the fuel to burn properly.

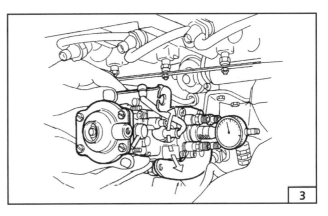

3. Unless the fuel pump, or the drive to the fuel pump, has been disturbed, it is unlikely that the fuel pump timing will be incorrect, especially if the engine has been running satisfactorily previously. However a worn drive belt can affect timing, while pumps do eventually wear and if you suspect an incorrectly set or faulty injection pump, it must be checked, either statically or (preferably) dynamically, with the appropriate test equipment. See **Appendix, Tools and Equipment** for examples of the test equipment needed.

The manufacturer's manual will give details on the correct method for setting the injection pump drive. Usually, irrespective of whether the drive is via belts, chains or gears, there are marks on the appropriate drive wheels which should be aligned with each other.

IMPORTANT NOTE: Turning the crankshaft quickly, (for example, by the starter motor), while the drive to the camshaft is disconnected, invariably results in the pistons striking the valves and consequential internal engine damage. If it is necessary to turn the crankshaft, remove the glow plugs, turn the engine gently by hand in the normal direction of rotation, and stop if any resistance is felt. Once you have reconnected the camshaft and injection pump drive, ensure there is no chance of damaging the valves and pistons. Turn the engine over at least twice by hand, stopping if any resistance is felt.

There is a possibility, albeit a remote one, that the timing may be out because a drive wheel has rotated on its shaft. Drive wheels and pulleys each normally have a key which fits into a keyway on the shaft and another in the drive wheel. Commonly, the key is a half-moon (Woodruff) type but they do come in all shapes and sizes. If all else fails to solve the black smoke it's worth checking the condition - and presence! - of the key, just to make sure.

IMPORTANT NOTE: In normal running it is unusual for the injection timing to suddenly become incorrect so it is more likely that the settings were wrongly set during reassembly or maintenance.

Check 4. Advanced fuel injection pump timing.

Advanced injection pump timing will result in fuel being injected before the optimum point. The combustion process may be erratic and the fuel will not burn completely.

The comments relating to retarded pump timing apply just as much to advanced pump timing. However, some pumps have a cold-idle advance device which may be mechanically or electrically operated. Check to make sure that this device is working properly and is not advancing the pump timing when the engine has warmed up. See **Appendix, Tools and Equipment** for brief details on cold-idle advance checking.

Check 5. Poor cylinder compressions.

Poor cylinder compressions mean that not all the air that the engine is theoretically capable of using is being compressed. The problem is usually leakage past worn piston rings or badly seating valves. However, any other fault which allows leakage, such as a leaking head gasket, or reduces the amount of air entering the cylinder, such as a restriction in the induction system, can cause poor compressions. If the engine is worn, it will have to be overhauled.

A cylinder compression tester will show up low compressions and which cylinders are affected. In the hands of an experienced operator, a cylinder leakage tester will give more information on the problem. Only use compression testers that are intended for diesel engines and follow the maker's instructions, particularly those relating to safety. See **Appendix, Tools and Equipment** for details of both testers.

Check 6. Sticking injector.

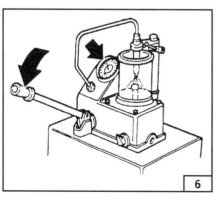

6. Unless they are very old and worn, contaminated with dirt, or the engine has overheated, sticking injectors are unlikely. Apart from checking to see whether the injectors produce a fine spray pattern there is little else that can be done, since a thorough examination requires specialist test and cleaning equipment. See **Appendix, Tools and Equipment**.

Check 7. Poor boost pressure (turbocharged engines).

Specialist test equipment is normally needed to properly diagnose turbocharger problems, although on most engines it is possible to check and see whether the wastegate is sticking open. In general though, it is not recommended that you try to make adjustments to the turbocharger control system unless you have access to the necessary equipment and manufacturer's data. The equipment and expertise necessary for diagnosing turbocharger problems is available at diesel engine specialists, such as members of the Perkins Powercentre network.

SAFETY FIRST!

• *A turbocharger becomes extremely hot in use, so let it cool down before attempting any work in or around the turbocharger area.*

Check 8. Incorrect valve clearances.

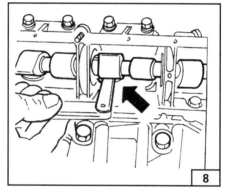

8. If the clearance between the end of the valve stem and the rocker arm (or camshaft lobe on OHC engines) is too small, the valves may not seat properly, which will result in poor compression. Some of the compressed fresh air leaks out lowering the compression pressure in that particular cylinder.

Too large a clearance will prevent the valves from fully opening, reducing the amount of air sucked into the cylinder during the induction stroke and preventing all the exhaust gases escaping during the exhaust stroke. Some modern engines may be fitted with hydraulic valve lifters which cannot be adjusted - although they can fail - so this is worth checking before assuming there is a problem with valve clearances.

Consult the manufacturer's manual for the correct valve clearances. It is usually essential that they are checked with the engine cold but some manufacturers may give clearances for hot engines. The manual should also indicate the correct order for checking the clearances.

If the engine has been rebuilt or the head gasket changed, the clearances should be checked before the engine is run. Traditional engines are re-checked after a short running time (specified by the manufacturer) but some modern engines - which do not need the head to be re-torqued down - do not need to be re-checked.

All non-hydraulic types must be checked and, if necessary, re-set at the manufacturer's recommended service intervals.

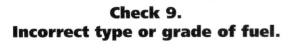
Check 9.
Incorrect type or grade of fuel.

The incorrect type or grade of fuel can cause incomplete or erratic combustion. Black smoke and high emissions are the result.

Obtain a small quantity of fuel that you know to be the right type and grade for your engine. Now set up an emergency fuel supply to the injection pump - see *Chapter 3, Part C*. Start the engine and run it. If the fuel in the tank is the wrong type or grade any black smoke will disappear as good fuel flows to the injectors.

Check 10. Faulty cold starting equipment.

The most common type of cold start aid is the glow plug, an essential aid for indirect injection engines. Faults with glow plugs - either failure to operate or to switch off - should not cause black smoke. Some cold start systems, such as the Perkins Thermostart system, burn fuel in the inlet manifold. These may leak fuel into the manifold and it is this excess fuel which causes black smoke.

The electric and fuel connections to the Thermostart system should be checked and the unit removed for physical examination.

FAULT 2: ENGINE PRODUCES WHITE OR GREYISH WHITE SMOKE

Check 11. Cold running.

If an engine runs at a lower temperature than intended, unburned fuel and oil vapour may pass out through the exhaust, characterised by white or greyish white smoke. Cold running is normally due to a faulty thermostat.

Remove the thermostat housing and check to see if a thermostat is fitted. On some larger engines there may be more than one thermostat - such as one for each bank of a Vee-type engine - and there is sometimes more than one thermostat fitted into a housing.

If a thermostat is there, remove it and check the specification. Various numbers are often stamped on thermostats and one of these will be the opening temperature. The manufacturer's manual should tell you the intended opening temperature which may be different for summer and winter. Test the thermostat by putting it in a pan of warm water along with a suitable thermometer. Heat the water up and note the temperature when the thermostat starts to open. This should be within a few degrees of the figure stamped on it.

Thermostats are reliable and do not often give trouble. When they go wrong they cannot be repaired and have to be replaced.

Check 12.
Retarded fuel injection pump

See *Check 3*.

Check 13.
Leaking injector or injectors.

The fuel from a leaking injector will vaporise inside the hot cylinder and pass out the exhaust as white smoke. See *Check 6*.

FAULT 3: ENGINE PRODUCES BLUE OR GREYISH BLUE SMOKE

Check 14. Leaking valve seals or worn valve guides & valve stems.

Worn valve stems, valve stem seals or valve guide bores can allow oil to migrate into the combustion chambers where it burns, most commonly producing blue smoke. Some dismantling, including cylinder head removal, is necessary to check valve stems and valve guides.

Check 15. Worn piston rings or cylinder bores.

Worn piston rings and cylinder bores will result in leakage of oil into the combustion chambers and low cylinder compressions. Broken or sticking piston rings can have a similar effect. A compression test, or better still a cylinder leak-down test, may be able to pinpoint the problem. See *Appendix, Tools and Equipment*. Confirming this fault will require some dismantling since low compressions can also be caused by worn or damaged pistons, badly seating or sticking valves and leaking head gaskets. An overhaul is the only cure. See *Chapter 3, Part C*.

Check 16.
Leaking turbocharger oils seals.

Turbochargers are subjected to extremely arduous operating conditions - cool intake air on one side, and - very hot exhaust gas on the other side - all while turning at many thousands of revolutions per minute. The lubricating oil and oil seals are worked very hard.

Remove the intake ducting at the inlet to the turbocharger and examine the internal surfaces of the intake passages for any oil contamination. Next, remove the exhaust pipe at the turbocharger. Check for any oily or greasy deposits. If any oil leaks are suspected, the turbocharger will have to be removed for specialist overhaul. See *Check 7*.

Check 17. Lubricating oil too thin.

Using lubricating oil that is too thin is a possible cause of both white and blue smoke. The thin oil can escape up past piston rings and down through valve guides into the combustion chambers. Without an expensive chemical analysis, there is no definite way of checking the type of oil in the sump and a badly worn engine is a more probable cause of oil leakage into the combustion chambers. If you are convinced the engine is in good condition, changing the oil and filter is the easiest way of making sure the correct grade of lubricant is in the sump.

FAULT 4: EXCESSIVE SMOKE OR VAPOUR FROM CRANKCASE VENTILATION OUTLET

Check 18. Worn piston rings or cylinder bores.

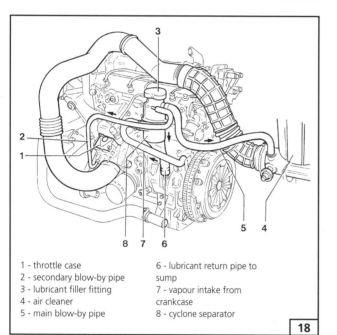

1 - throttle case
2 - secondary blow-by pipe
3 - lubricant filler fitting
4 - air cleaner
5 - main blow-by pipe
6 - lubricant return pipe to sump
7 - vapour intake from crankcase
8 - cyclone separator

18

At the same time as oil can seep past pistons into the combustion chambers, high pressure gases can leak past the piston rings into the crankcase. The hot gases increase the pressure in the crankcase and mix with the oil vapours - an excessive flow of oily smoke and vapour from the crankcase ventilation outlet is the result.

18. This is more apparent on engines that vent to the atmosphere. On engines that 'recirculate' the crankcase gases into the inlet or exhaust manifolds, remove a pipe and check the condition of the crankcase ventilation system. Pipes can collapse internally, or be trapped or kinked, or the system may become blocked, causing oil to be forced out of gaskets and seals. A compression or leak-down check will help to confirm whether the piston rings or cylinder bores are worn. See *Check 5*.

Check 19. Defective exhauster (if fitted).

Since a diesel engine's inlet manifold vacuum is minimal, some other way has to be found to provide the vacuum needed for servo assisted automotive brake systems. Various types of vacuum pump are used but one type, called an exhauster, vents into the crankcase.

It is possible that a leaking vacuum system will allow an exhauster to pump air into the crankcase, although the lack of effective servo-assistance to the brakes will probably be noticeable long before the outlet from the crankcase breather is checked.

The increased flow of air through the crankcase picks up oil mist and droplets which can come out of the breather outlet as an oily vapour. If the engine recirculates the crankcase breather into the inlet or exhaust manifolds, this increased flow of oil can also cause the engine to smoke. Eventually the breather system can become blocked resulting in excessive crankcase pressures and oil being forced out of gaskets and seals. The vacuum system will have to be overhauled and the faulty components replaced.

PART C: REPAIRS AND MAINTENANCE.

FAULT 1: ENGINE PRODUCES BLACK (OR DARK GREY) SMOKE

Job 1. Induction system restriction.

Apart from a blocked air filter, a restriction in the induction system is unlikely unless, of course, the engine has an air shut-off valve.

❑ **Step 1.** If the engine has an air shut-off valve, check to see whether it is fully or partly closed. If necessary, reset it by moving the manual reset lever from the CLOSED to the OPEN position. Some types of emergency valve, such as those supplied by Chalwyn Equipment, should reset automatically when the engine stops, but do check to make sure this has happened.

SAFETY FIRST!

- *Find out why an air shut-off valve was triggered.*
- *Maybe there was a good reason and continuing to run the engine could be dangerous or cause damage.*
- *Always find out the facts about past problems before continuing to run the engine, and always exercise extreme caution until the source of the problem is found.*

Step 2. If the air shut-off valve is open (or your engine does not have one), disconnect the air filter inlet at the manifold and run the engine. If the black smoke disappears, you have a clogged air filter or an obstruction in the air inlet ducting.

Step 3. Examine the air filter, looking for any obvious dirt or dampness. The only sure way of eliminating the filter as a cause of the intake restriction is replacement or cleaning. There is a very wide range of filter types and installations so you should consult the manufacturer's manual for specific instructions. See **Chapter 4, Part C, Job 7** for general details.

Step 4. Look carefully at the intake ducting for any damage and dismantle it sufficiently to be able to see inside. Any damaged or kinked ducting should be repaired or replaced, and any blockages should be removed. Try to find out what caused the damage or blockage so you can prevent it from happening again.

Job 2. Dirty, worn or incorrect injectors.

It is unlikely that all the injectors on a multi-cylinder engine will develop faults at the same time. However, it is possible to fit a set of incorrect injectors - many injectors look the same but the internal components and settings vary to suit different engines.

Step 1. Check all the injectors to make sure they match the manufacturer's specifications.

Step 2. *In-situ* injector testers are available which will allow setting pressure and back leakage to be checked without removing the injectors from the engine. See **Appendix, Tools and Equipment**. An experienced operator will also be able to get some idea of an injector's condition by watching the behaviour of the gauge.

SAFETY FIRST!

- *The high pressure fuel spray from an injector can penetrate the skin, and allow fuel to enter the bloodstream, possibly with fatal consequences.*
- *Only operate an injector while in the engine or while in a proprietary, enclosed injector tester.*
- *Atomised diesel fuel can burn quite easily so do not use any naked flames near a spray.*

Step 3. Disconnect the high-pressure pipes and remove the injectors. Reconnect the injectors to a fully-enclosed high pressure tester and watch the fuel spray while the injector is worked. Anything other than an even and regular spray indicates a problem. This is a very basic check which will only show up very advanced injector deterioration.

Step 4. If any injectors are suspect, fit new or serviceable used replacements, making sure that they match the manufacturer's specifications. Any spacers or insulating washers should be replaced in accordance with the manufacturer's instructions.

Step 5. Injectors are manufactured to very fine tolerances and need specialist equipment to accurately check their condition and make setting adjustments. Unless users have access to this equipment, attempts at repair or overhaul are not recommended. Injectors should be returned to the manufacturer or distributor, such as a member of the Perkins Powercentre organisation, for an assessment of their condition and a judgement on whether they are repairable.

Step 6. With access to the right test equipment such as the tester shown in **Appendix, Tools and Equipment**, injectors can be tested, dismantled for cleaning and, assuming components are within tolerance, reassembled. Some items, such as springs and shims can be replaced but if the injector plunger and body show signs of wear they must be scrapped.

i INSIDE INFORMATION: An intermittently sticking injector can be very hard to find and substitution with a new or serviceable used injector may be the only way of completely eliminating this fault. *i*

Job 3. Retarded fuel injection pump timing.

Find out whether any recent work has been carried out on the engine, such as pump replacement or renewal of the pump drive. It is very unlikely that a correctly set pump will suddenly lose its setting and problems with pump timing are more likely to be due to incorrect setting after related work, or fitting a similar-looking but differently-specified pump. If the pump has not been disturbed for many years, cumulative wear either within the pump or in the drive mechanism is a likely cause of incorrect timing.

Step 1. Check the pump and make sure it is the one specified for the engine. Replace an incorrect pump with the correct component. See **Chapter 3, Part C**.

Step 2. If there are any doubts about a pump's serviceability, it should be removed for testing. During removal, take note of the way any linkages or pipes are connected and any shims used to adjust the pump timing. All openings must be sealed to prevent dirt getting into the pump. Sophisticated test equipment is needed to assess a pump's condition and this work is best left to a specialist, such as member of the Perkins Powercentre network.

Step 3. Refit the pump, if removed, and check the pump timing. See **Chapter 4, Part C, Job 19** for details on pump timing.

Job 4. Advanced fuel injection pump timing.

The comments in **Job 3** about retarded injection pump timing apply just as much to advanced pump timing. However, some engines are fitted with cold-idle advance mechanism which advance the injection, usually by 4 to 6 degrees, when the engine is cold. Before assuming the basic pump timing is at fault, check to make sure any cold-idle advance devices are working properly.

Brief details for using a diesel engine timing tester are give in **Appendix, Tools and Equipment**. The testing and repair procedures vary, depending on whether the engine has a mechanical or all-electric cold-idle advance system.

PART A: MECHANICAL COLD-IDLE ADVANCE

☐ **Step A1.** The injection pump mechanism test can be carried out with the engine cold, or at normal operating temperature but the cylinder head temperature sender test should be carried out as the engine warms up from cold.

☐ **Step A2.** Prepare and connect a diesel engine timing tester in accordance with manufacturer's instructions. Check the engine manufacturer's manual for the correct cold-idle advance settings.

☐ **Step A3.** Disconnect the linkage from the cold-idle advance device (arrowed), start the engine and manually operate the cold-idle advance.

Job 4-A3

☐ **Step A4.** Note the dynamic timing figures as the cold idle advance is operated. A variation of 4 to 6 degrees is normal.

☐ **Step A5.** If no change is seen, the internal pump mechanism is faulty and the pump should be removed for overhaul.

☐ **Step A6.** If the pump mechanism appears to be working correctly, watch the operating cable as the engine warms up. If the cable moves, the sender is probably

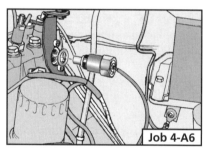
Job 4-A6

working correctly. If there is no movement the sender should be replaced. Consult the manufacturer's manual for the correct setting figures.

PART B: ELECTRICALLY OPERATED COLD ADVANCE

Before carrying out the following checks, make sure that:
• all electrical terminals are sound
• the earth/ground connection is effective

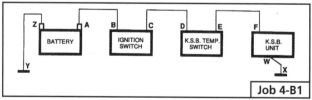
Job 4-B1

☐ **Step B1.** The Bosch KSB system is used here as an example. Make sure the engine is cold (coolant below 20 degrees C). With the ignition off, connect a voltmeter between the relevant terminals - see the manufacturer's

manual. A zero voltage is correct - anything else and the electrical system is faulty. Check back through the system to locate the fault. Repair or replace any suspect components.

☐ **Step B2.** Turn on the ignition and check the voltage. Again, a zero voltage is correct. If battery voltage is indicated, the temperature switch is faulty or there is a wiring fault. Check back through the electrical system to locate the problem. Repair or replace any suspect components.

☐ **Step B3.** Disconnect the fuel cut-off solenoid to prevent the engine from starting. With the ignition on, crank the engine. Again, a zero voltage is correct. If battery voltage is indicated, the temperature switch is faulty or there is a wiring fault. Check back through the electrical system to locate the problem. Repair or replace any suspect components.

☐ **Step B4.** Reconnect the fuel cut-off solenoid and disconnect the unit at the relevant connection - see the manufacturer's manual. Start the engine and allow it to warm up. At about 40 degrees C or above, the voltmeter should show battery voltage. If the reading is zero volts, check back through the electrical system to locate the problem. Repair or replace any suspect components.

☐ **Step B5.** Listen to the engine note while the engine is revved then reconnect the cable to the unit and wait about three minutes. Rev the engine again and listen to the engine note. If the engine is less noisy and any smoke has disappeared, the unit is working. If there is no change, the unit is faulty, and specialist advice from an agent dealing with the unit should be sought.

Job 5. Poor cylinder compressions.

☐ **Step 1.** Test with a diesel engine compression tester. See *Chapter 3, Part C*.

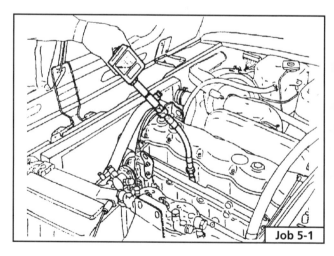
Job 5-1

Job 6. Sticking injector.

See *Job 2*.

Job 7. Poor boost pressure (turbocharged engines).

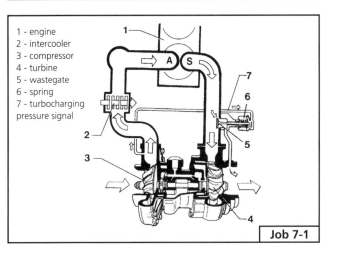

1 - engine
2 - intercooler
3 - compressor
4 - turbine
5 - wastegate
6 - spring
7 - turbocharging pressure signal

Job 7-1

❑ **Step 1.** The specialist test equipment needed to properly diagnose and repair turbocharger problems is available at manufacturer's dealers or support centres. However, a simple check is to make sure that the wastegate is not sticking open. Designs vary, so check the manufacturer's manual for details on wastegate location. It is usually necessary to dismantle part of the exhaust system in order to gain access to the wastegate.

Job 8. Incorrect valve clearances.

IMPORTANT NOTE: Before assuming that the valve clearances are incorrect, check to see whether your engine has hydraulic valve lifters. This type cannot be adjusted - although they can fail. Noisy hydraulic lifters will have to be renewed.

Apart from causing smoke and a build up of deposits in the engine, continuing to run an engine with incorrect valve clearances can cause irreparable damage. Too small a clearance and the valves, particularly the exhaust valves, may melt. Excessively large clearances and the sudden loads on the valve train may cause premature wear.

See **Chapter 4, Part C, Job 16** for details on setting valve clearances.

Job 9. Incorrect type or grade of fuel.

Once you're sure the fuel in the tank is wrong, the decision has to be made on how to get it out and where to put it. This will depend on the type and location of the tank and the quantity of fuel. You have the choice of pumping the fuel out, siphoning or draining it. Pumping will work in most situations, while siphoning and draining will only work if the receiving container is lower than the tank. Do not worry about getting every last drop of fuel out. If you completely refill the tank, any remaining old fuel will be well diluted.

❑ **Step 1.** Obtain a small quantity, around five litres, of known good fuel and set up an emergency fuel supply. See

Chapter 3, Part C. Preferably, the supply should feed into the fuel filters, but not much harm will be done if it feeds direct to the injection pump, provided the supply is perfectly clean.

❑ **Step 2.** Run the engine until you are sure that the good fuel is getting to the injectors. If the black smoke does not stop, it is being caused by another fault.

❑ **Step 3.** If the black smoke stops, drain or siphon the contents of the fuel tank into suitable containers and refill the tank with good fuel. Dispose of the old fuel safely, at your local authority disposal amenity.

> **ENVIRONMENT FIRST!**
>
> • Dispose of the waste fuel at a proper site.
> • Do not pour it into drains, waterways or anywhere it can seep into the river system.

Job 10. Faulty cold starting equipment.

The CAV Thermostart cold-starting system burns fuel in the inlet manifold to pre-heat the intake air. If the fuel valve leaks, the system will cause over-fuelling.

❑ **Step 1.** Disconnect the fuel supply to the Thermostart system. If this cures the problem, there is a fault in the Thermostart unit which should be overhauled or replaced.

❑ **Step 2.** Check the electrical supply to the Thermostart unit and make sure it is switching off at the correct time.

FAULT 2: ENGINE PRODUCES WHITE OR GREYISH WHITE SMOKE

Job 11. Cold running.

❑ **Step 1.** Remove the coolant thermostat housing and check to see whether a thermostat is fitted. If one (or more, as appropriate) is fitted, check to make sure it is the correct type and specification, and that it is working properly. See **Chapter 4, Part B, Check 87** for thermostat testing.

❑ **Step 2.** Fit a new thermostat or refit the existing thermostat if it seems serviceable.

Job 12. Retarded fuel injector pump timing.

See **Job 3**.

Job 13. Leaking injector or injectors.

See **Job 2**.

FAULT 3: ENGINE PRODUCES BLUE OR GREYISH BLUE SMOKE

Job 14. Leaking valve seals or worn valve guides & valve stems.

Although most valves are fitted with oil seals, worn valve stems or valve guide bores can reduce their effectiveness.

☐ **Step 1.** Check the manufacturer's manual to see if any special tools are needed. Remove the rocker cover and rocker assembly.

☐ **Step 2.** Worn valve stems or valve guide bores can only be confirmed by dismantling. At the very least the valve springs will have to be removed, which can be done with the cylinder head in place. See **Appendix, Tools and Equipment** for the correct type of valve spring compressor to allow valve spring removal without removing the cylinder head. If there is any perceptible side-to-side movement once the springs have been removed, the wear is excessive.

☐ **Step 3.** Decide which cylinder to work on and turn the engine until the piston is at TDC. Use a lever-type compressor to compress the springs. Remove the collets, springs and old seals. Keep a note of any seats or spacers. Wear suitable eye protection when removing collets.

making it easy!
- Valve collets are small and can easily fall down many of the openings in the cylinder head.
- Use rags or paper wipes to block these holes before compressing the valve springs.
- Grease can be used to hold the collets in place as the valve springs are decompressed during reassembly.

☐ **Step 4.** If no wear is apparent, fit new valve seals and, preferably, new springs. Repeat the process for each cylinder

☐ **Step 5.** If necessary, remove the cylinder head and dismantle the valves. Use a micrometer to measure the valve stem diameters and compare the figures with the manufacturer's specifications. If there is any wear beyond acceptable tolerances, they will have to be replaced.

☐ **Step 6.** See **Chapter 4, Part C, Job 18** for details on replacing valve guides.

☐ **Step 7.** Reassemble the cylinder head and refit to the engine using a new gasket. Torque the head bolts in the correct sequence to the figure specified by the manufacturer.

☐ **Step 8.** Reassemble the rocker gear, check the valve clearances and fit the rocker cover.

Job 15. Worn piston rings or cylinder bores.

These faults can only be confirmed by dismantling the engine and examining the condition of the piston rings and bores. If no damage or wear is apparent, accurate measurement will confirm whether the components are serviceable. See **Chapter 3, Part C**.

Job 16. Leaking turbocharger oil seals.

See **Job 7**.

Job 17. Lubricating oil too thin.

> **SAFETY FIRST!**
> • *Hot engine oil can scald so let a hot engine cool down before draining the oil and removing the filter.*

☐ **Step 1.** Drain the oil and remove the old filter.

> **ENVIRONMENT FIRST!**
> • *Old engine oil should be disposed of properly and not, under any circumstances, dumped into drains.*
> • *Local authorities often provide collection tanks at their waste disposal sites.*

☐ **Step 2.** Fit a new filter or filter element and refill the engine with the correct grade of oil.

FAULT 4: EXCESSIVE SMOKE OR VAPOUR FROM CRANKCASE VENTILATION OUTLET

Job 19. Defective exhauster (if fitted).

Check the vacuum system for any leaks and repair or replace any leaking components. Details for servicing a typical exhauster are given in **Chapter 6, Ancillaries**.

CHAPTER 6
ANCILLARIES

Diesel engines are used in an enormous variety of applications.

Although their size and power varies, there are certain components and ancillaries that are common to every engine, such as fuel injection pumps, injectors and filters. In addition, many other ancillaries such as hydraulic pumps, air compressors and safety valves are used with, or fitted to diesel engines, to suit specific applications.

Marine diesels also have special cooling and exhaust equipment. Fuel pumps, injectors and many other components have been mentioned previously in some detail. This Chapter examines some of the ancillaries that have only had a very brief mention, as well as the needs of marine cooling systems, giving some idea of how they operate and what can be done to reduce the chances of things going wrong.

Chapter Contents

Filters

Clean air, oil and fuel are essential for long engine life, so making sure that the right filters are used and serviced at the appropriate times is a priority. Filters are cheap compared with the cost of an engine rebuild.

There are almost as many filter types as there are engines but there are really only two general types: barrier and centrifugal. Most air filters are either the barrier type or the barrier and centrifugal combined.

You may consider that the oil bath filter is a third type but it can be considered a variation of the centrifugal filter and, since it is normally used with an oiled wire mesh element, it is a combination of the two types. Fuel filters, along with most air filters, are of the barrier type but centrifugal oil filters do exist.

In the barrier type of filter, the fluid passes through a porous barrier which is often paper but can be fabric, sintered metal or, in older air filter systems, an oiled wire mesh.

Whichever particular type is used, the aim of the barrier is to trap the contamination while allowing the air or liquid to pass through.

AIR CLEANERS AND FILTERS

1. When combined with barrier elements, centrifugal filters such as this two stage 'Cyclone' type can be very efficient. The intake air is made to move in a spiral path and the heavier contaminant particles are thrown out

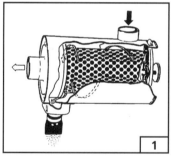

against the walls of the filter case either to be collected or, as in this case, ejected through a rubber valve. These filters are carefully designed to give the air the necessary swirl and should not be modified.

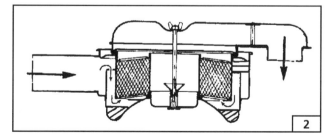

2. With an oil bath cleaner the air is drawn into the cleaner and directed to the oil surface where it changes direction to flow into the wire mesh. Although the air can easily change direction the heavier contaminant particles go straight on into the oil. The oil level is critical: too low, and it's easier for contaminants to get into the engine, while too high and oil may be picked up and sucked in with the air, leading to uncontrolled engine speed and excessive engine wear.

OIL BATH AIR CLEANER

3. This is another common type of oil bath air cleaner, with a centrifugal pre-cleaner, suitable for removing coarse particles and useful in more hostile environments.

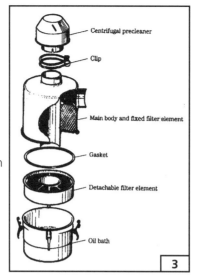

Centrifugal precleaner
Clip
Main body and fixed filter element
Gasket
Detachable filter element
Oil bath

3

4

4. Remove the lid and lift out the element. Drain the oil from the container and clean out the dirt and sludge. Refill the container with fresh engine lubricating oil, to the indicated level. Clean the element in paraffin (kerosene) and, depending on the manufacturer's recommendation, either replace it dry, or dip it in oil and allow it to drain before replacing it in the container. Refit the lid, making sure that the lid seats properly on the seal.

PAPER ELEMENT FILTERS

5. Undo the cover and remove the element. If the element appears clogged or dirty in any way, renew it. Clean out the element container making sure no dirt gets into the intake ducting. Wet dust, such as oil contamination cannot be removed and the element should be replaced. Make sure the element seats correctly and any rubber sealing rings are replaced, as necessary.

5

CENTRIFUGAL TYPE WITH ELEMENT

6. The Dry-Type Two Stage 'Cyclopac' is used as an example of dismantling. Unclamp the dust bowl (**6**), remove the baffle plate (**5**) and clean out the bowl. The dust in the bowl should not be allowed to reach within 1/2 in. (13 mm) of the dust entry slot in the baffle. Release the wing nut (**3**) and remove the baffle. Dry dust can be removed from the element (**2**) by blowing back from the clean side of the pleats using air pressure less than 100 lb/in2 (7 kgf/cm2). If the element is contaminated by oil and/or soot, it can be cleaned in warm water using a suitable non-foaming detergent - check with the air cleaner manufacturer.

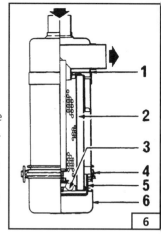

1
2
3
4
5
6

6

FACT FILE: SERVICING A REPLACEABLE ELEMENT

• Replace any element which is damaged or which cannot be satisfactorily cleaned by the method described.

• Never exceed recommended maximum pressures when using compressed air to clean the element.

• Never strike the element against a hard surface to dislodge the dust. This will damage the sealing surface and may also rupture the element.

• Never 'blow' dirt out of the filter housing. This may introduce dust into the engine. Instead, use a clean, damp cloth.

• Do not oil the element.

• Always use eye protection when using compressed air.

ELEMENT INSPECTION

Very large elements can be very expensive to replace. Vacuum cleaning the 'dirty side' of the element will remove much of the accumulated dust. However, this is not considered an adequate cleaning procedure, and the element should be further serviced as follows:

• 'Reverse flush' the element with a stream of clean, oil-free compressed air reduced to less than 30 lb/in2 (2 kg/cm2) with effective personal protective equipment. Direct the stream up and down the pleats, opposite the normal air flow from the 'clean air side' of the element. Continue this 'reverse flushing' procedure until all dust is removed.

• If any oil or exhaust soot remains on the element and creates unwanted air restriction, it will be necessary to replace the element with a new one.

7. • When the element is satisfactorily cleaned, inspect it for damage before placing it into full service (See **ELEMENT INSPECTION on page 6-3**).

USING BRIGHT LIGHT INSIDE OF ELEMENT TO INSPECT FOR LEAKS OR DAMAGE

7

ELEMENT INSPECTION

Inspect the cleaned element, if of the larger type, by placing a bright light inside and looking through the element. If there are any thin spots, pinholes or other damage, the element should be replaced. The element should be renewed after six cleanings, or once a year, whichever comes first. Clean the inside of the filter body and fins, making sure that no dirt enters the air filter outlet. Check all hoses for condition and security. Reassemble the air cleaner unit.

OIL FILTERS

Normally, only barrier type filters are used, but a centrifugal type could be fitted . There are two basic styles of barrier type oil filter: the replacement element type and the canister (or throwaway) type. Although many canister oil filters may appear similar from the outside, they can be quite different inside, so always use the filter specified for your engine.

REPLACEMENT ELEMENT FILTERS

8. Replacement element filters, mainly found on old engines, are generally easy to remove, being held in place by a long central bolt. Be ready to catch any oil spillage as the filter body is loosened and take note of any spacers, washers, springs and gaskets that may be fitted inside the holder.

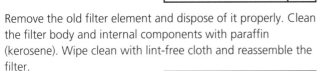

Remove the old filter element and dispose of it properly. Clean the filter body and internal components with paraffin (kerosene). Wipe clean with lint-free cloth and reassemble the filter.

CANISTER FILTERS

9. Using a suitable tool, such as a strap wrench, canister filters are easy to remove. Spin off the old filter, while being ready to catch any oil spillage.

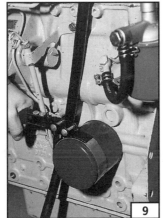

10. Clean the filter housing, smear some clean engine oil on the rubber sealing gasket and spin on the new filter. Tighten the filter up to 3/4 of a turn - by hand only.

CENTRIFUGAL OIL FILTERS

11. Thoroughly clean all rotating components. Failure to do so may result in components running out of balance, which would accelerate spindle wear. Examine bearings for wear and replace with an approved replacement part if excessive wear is suspected. Reassemble, making sure that all rotating parts are free to move and all covers and clamps are securely in place. Prime the filter with the special pump, start the engine and check for leakage or excessive vibration.

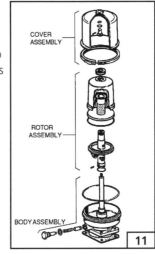

FACT FILE: FUEL FILTERS

12A. There are three main designs of diesel fuel filter. The first has a removable canister (**a**) with a disposable inner element (**b**) and sealing ring (**c**).

12B. Another type has a disposable filter body (**a**) sandwiched between water bowl (**b**) and filter head (**c**).

12C. The third type has a fully disposable filter cartridge (**a**), attached to the pump and oil cooler connections on this Ford unit.

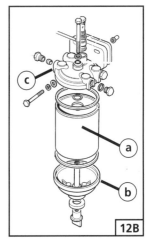

12D. If your engine is fitted with a water level sensor in the filter, wired to a tell-tale lamp, you can afford to leave water-drainage for longer. Engines with this type of Lucas filter but no such sensor, can have one retrospectively fitted.

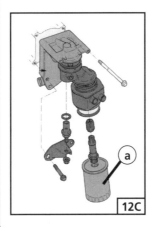

CONTINUOUS ENGINE RUNNING

13. Some diesel engines run continuously and stopping them could be very inconvenient or dangerous. For example, generating sets in remote areas or some marine

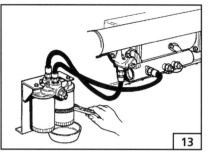

diesels. These engines are often fitted with change-over oil and fuel filters that can be removed and replaced with the engine running. An engine fitted with change-over filters will have pairs of filters (oil, fuel or both) and a valve that switches the supply from one to the other or, for normal running, through both. You should consult the manufacturer's manual for detailed instructions on removing and replacing change-over filters. The techniques are the same for both oil and fuel filters, except that with some fuel filters, the change-over valve must only be operated with the bleed valve open, until all air has been ejected.

Compressors and Pumps

Depending on the application, engines may be fitted with air compressors, hydraulic pumps, power steering pumps, bilge pumps or vacuum pumps. The detailed operation of any of these components is beyond the scope of this book but there are general considerations which should be borne in mind.

14. Manufacturers' manuals should be consulted for details of pump and compressor components, such as of the bilge pump shown here. The very basics of any fault finding programme should include inspections to ensure that:
• mountings are secure and undamaged,
• belts are at the correct tension,
• hydraulic reservoirs are at the correct levels,
• valves and especially any safety valves on compressors or hydraulic pumps are working properly,
• connecting pipes are undamaged and connections are secure
• bearings should be checked for wear with drive belts slackened.

Air Shut Off Valves

15. Air shut-off valves are extra safety features that can be found on some diesel engines. Operation can be manual or automatic, the automatic shut-down being triggered either by engine overspeed, low oil pressure or both. These valves can be fitted into the intake ducting or can be combined with air filters. The details

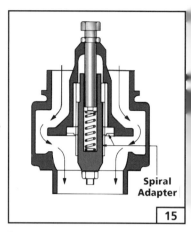

mentioned here relate to valves supplied by Chalwyn Equipment. Other manufacturer's requirements may differ.

OPERATING PRINCIPLE

An air shut-off valve is essentially a spring loaded poppet valve fitted upstream of the inlet manifold or combined with an air filter. The valve is operated by the pressure difference across it when the engine is running. As engine speed increases the air pressure on the valve tries to close it but this is resisted by springs. However, the tension in the springs can be adjusted so that the valve closes when the engine reaches its maximum governed speed. Once closed, the valve is held against the seat by the partial vacuum in the manifold. As the vacuum leaks away, the springs reset the valve.

16. A cable operated manual override may be fitted to the valve to give the operator control over shutdown. The automatic shut down on overspeed still functions and the manual override can be attached to other pneumatic or hydraulic

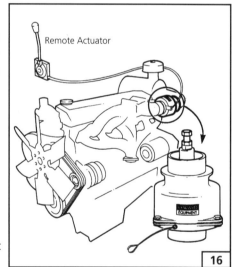

actuators. Taking things a stage further, air shut off valves can be fitted with flame traps and automatic shut down if lubricating oil pressure falls dangerously low.

The airflow direction through the valve should be indicated by arrows on the valve body. Valves may be fitted at any angle. Ideally valves should be fitted near the inlet manifold and connected by rubber hoses and clips. If the valve has been fitted remote from the manifold, it should be connected by metal pipes or metal-reinforced-rubber. On turbocharged engines it is normally only possible to fit the valve upstream of the turbocharger. Any existing breather connections to the manifold should be re-routed to a position upstream of the valve.

FACT FILE: ADJUSTING THE VALVE

17. When supplied, each valve will be set to give a shutdown speed well below the maximum governed speed. Operators have to make the final adjustment themselves. Note the speed the engine shuts down and then, following the manufacturer's instructions make small adjustments to increase the speed until maximum governed speed is reached. The exact adjusting procedure varies with different valve designs. The following procedure relates only to the Chalwyn Spiral Type and is given as a guide to the necessary adjustments.

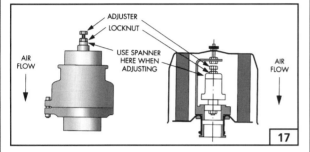

Where excessive wear or damage has occurred, the valve should be returned to the manufacturer for appraisal and reconditioning.

- a. Loosen locknut.
- b. Turn adjuster clockwise.
- c. Tighten locknut.
- d. Start engine - note shutdown speed.
- e. Repeat steps **a** to **d** until maximum governed speed is reached without shut down.
- f. Stop engine and apply any final overspeed adjustment - use thread lock adhesive on the locknut. If insufficient adjustment is available, obtain a stronger spring from the manufacturer.

IMPORTANT NOTE: When fitted to turbocharged engines, the valve should be adjusted when the engine is operating under full load.

SAFETY FIRST!

• In certain hazardous areas, regulations require a manual as well as an automatic shut down valve to be fitted. Consult the engine or safety valve manufacturer for details regarding installation and adjustment.

Exhaust Spark Arresters

Spark arresters enable diesel engines to operate in areas of fire hazard, by reducing the discharge of hot carbon particles through the exhaust pipe.

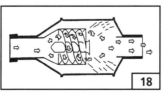

18. The stainless steel spiral causes the exhaust gases to rotate, throwing the hot carbon particles against the outer casing and cooling them before discharge. This type of spark arrester is suitable for fork lift trucks, towing tractors, etc.

SAFETY FIRST!

• When flammable gases are present in the atmosphere such as in coal mines, special spark arresters or flame traps must be fitted to both induction and exhaust systems to prevent explosion from engine induced sparks.

EXHAUST DILUTERS

19A. These are designed to mix large quantities of ambient air with the exhaust gas before releasing it to atmosphere.

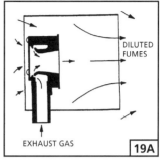

This cools and dilutes the exhaust gas constituents to levels that can be acceptable to personnel in the vicinity of the engine.

19B. The fine clearance of the nozzle gap may particularly affect back pressure in service, and periodic checks should be made to ensure that back pressure is still within the recommended limits.

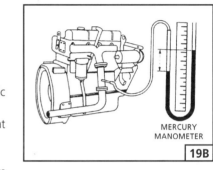

Automatic Overheat Protection

Many diesel engines run unattended and there is a possibility that a coolant system fault could result in an engine overheating and seizing.

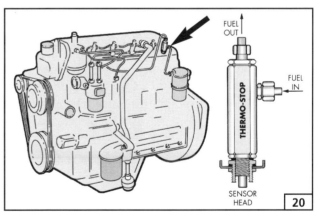

20. The Chalwyn 'Thermo-Stop' (arrowed) consists of a wax motor which is activated at a temperature of approximately 100 degrees C (212 degrees F) to close a valve which shuts off the fuel supply to the engine.

If 'Thermo-Stop' is activated, it can be reset once the engine has cooled down. Disconnect the outlet fuel pipe and insert a suitable metal rod through the end fitting. Push firmly to retract the wax motor to its pre-activated position. Reconnect the fuel pipe and bleed the system. Find out why the 'Thermo-Stop' activated before restarting the engine. If the valve cannot be reset, it cannot be repaired and will have to be replaced.

Automatic Low Oil Pressure Protection

21. Some diesel engines do not have oil pressure gauges and many run unattended.

A low oil pressure cut-off valve, such as the "Chalwyn Lube-Stop" automatically stops the engine if oil pressure falls to a dangerous value.

If the valve cuts the fuel supply, then cuts in again when the engine has cooled, check:
• coolant levels and cooling system efficiency
• oil level
• engine problems leading to overheating
• engine wear leading to severe drop in oil pressure when hot.

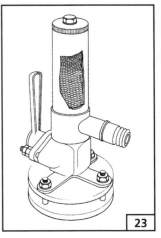

Marine Engine Cooling

The three types of water cooling used in marine engines are by direct sea water, by keel pipes and by fresh water and heat exchanger.

DIRECT COOLING

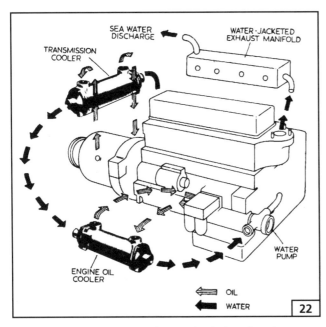

22. Direct cooling is the simplest method of cooling the engine but has its problems. Water is taken from outside the boat below the waterline, circulated around the engine, usually mixed with the exhaust gases and dumped over the side. The main problems are ensuring a clean water supply, an optimum coolant temperature and avoiding salt-water corrosion.

23. A strainer, often combined with a seacock, to isolate the strainer when it is being cleaned, is used to take the worst of the contaminants out of the water but a compromise has to be reached: Too fine a strainer and it is easily blocked; while too coarse a mesh and contaminants can enter the system, blocking the narrow passages within the engine.

This type of system should be checked frequently to make sure that the intake strainer is not blocked and that the sea cock is functioning correctly. Use a thermostat in the system to control coolant temperature - do not use the sea cock to control the flow. Removal of the thermostat can alter the flow of coolant around an engine, sometimes with detrimental effects such as local overheating.

KEEL COOLING

24. Keel cooling is a fresh-water closed-circuit cooling system and is a halfway house to full heat exchanger indirect cooling. The coolant water is circulated through pipes (arrowed) on the outside of the boat near the keel. This type of cooling system is unaffected by the cleanliness of the surrounding water and, because fresh water is used, the coolant temperature can be closer to optimum.

However, the coolant pipes on the outside are subject to marine fouling and may need quite frequent mechanical cleaning. They are also subject to damage and, if made from steel, to internal corrosion. Once a year the external skin pipes should be flushed out and their condition checked. In other respects, maintenance of this type of cooling system is similar to land based systems.

INDIRECT HEAT EXCHANGER COOLING

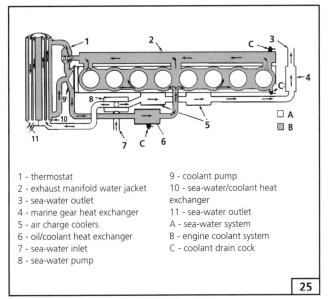

1 - thermostat
2 - exhaust manifold water jacket
3 - sea-water outlet
4 - marine gear heat exchanger
5 - air charge coolers
6 - oil/coolant heat exchanger
7 - sea-water inlet
8 - sea-water pump
9 - coolant pump
10 - sea-water/coolant heat exchanger
11 - sea-water outlet
A - sea-water system
B - engine coolant system
C - coolant drain cock

25. The indirect heat exchanger system operates on the same principle as the keel cooler but a special radiator (**10**) is fitted in the fresh water circuit instead of cooling pipes along the keel. This radiator is cooled by outside (raw) water pumped through it. In this respect it performs the same function as the radiator on a land based cooling system. Outside water is supplied to the heat exchanger via a sea cock and strainer, which should be regularly checked for blockage. The sea water pump is susceptible to damage from any contaminants passing through the strainer and most modern pumps are of the impeller type with rubber or neoprene vanes. These are quite resilient but vanes can break off and block internal passages, so check the pump at least once a season.

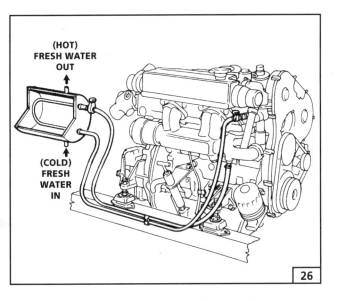

26. Sometimes, raw water is also used in an engine or gearbox oil cooler. Then, the hot coolant is often passed through an additional heat exchanger (or calorifier) to heat fresh water for cleaning and washing.

IMPORTANT NOTES: • Sea cocks and other parts of the cooling system are almost invariably below the water line so any leaks can have disastrous consequences, especially if the boat is left moored for a period of time. It is a good idea to close the sea cock whenever the engine is not required.

• Some turbo-charged engines have a strainer fitted to protect the intercooler. Remove, check and clean on a regular basis, if fitted.

Exhauster Vacuum Pump

There is no throttle in a diesel engine's air intake system so, unlike petrol engines, the vacuum in the inlet manifold is minimal. Alternative arrangements have to be made to provide the vacuum needed for servo assisted brakes and this is achieved with an engine driven vacuum pump, often called an exhauster.

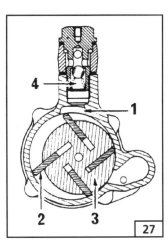

27. There are several different exhauster designs available, the type H.175 explained here being manufactured by Clayton Dewandre and used on some Perkins engines. The H.175 is a high speed rotary sliding vane unit with an eccentrically mounted rotor (**3**). Centrifugal force keeps the rotor blades (**2**) in contact with the bore. As a rotor blade passes the intake (**1**), the space between it and the following blade increases and air is drawn from the vacuum reservoir. Continued rotation decreases the space compressing the air which, along with any lubricating oil, discharged to the engine crankcase through the outlet. An oil check valve (**4**) prevents oil from passing into the vacuum pipe during operation.

Exhausters can be driven in several different ways, such as by flexible connections or belts. The type described here is gear driven. It is mounted on the rear of the timing case and is lubricated by engine oil.

SERVICING THE EXHAUSTER

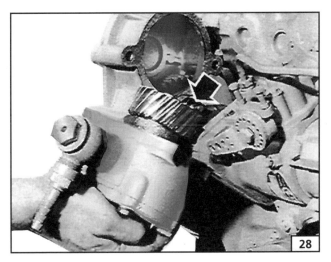

28. This type of exhauster should be removed, dismantled thoroughly cleaned and inspected every 60,000 miles or 2,400 running hours. However, other models have checks that should be carried out every 5,000 miles and 10,000 miles - the manufacturer's manual should be consulted.

After cleaning, the body should be examined for cracks and the body for scores or ripples, other components should be checked for wear. End float and running clearances should be within those specified by the manufacturer. If there is any wear or damage, components should be replaced, as should all the seals.

Reassemble and refit the exhauster using a new gasket and sealant if appropriate.

Marine Exhaust Systems

There are two general types of exhaust system used with smaller marine diesels; water cooled with or without a silencer, and the dry type. Whichever type is used it should be accessible throughout its length.

WATER COOLED EXHAUST

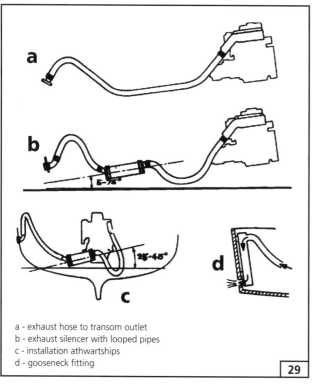

a - exhaust hose to transom outlet
b - exhaust silencer with looped pipes
c - installation athwartships
d - gooseneck fitting

29

29. Water cooled exhausts are used for two reasons: to ensure that the exhaust system runs cool enough to pass through enclosed areas of the boat without risk of fire, and to prevent temperatures in the engine compartment from reaching unacceptable levels. The first sign of a coolant blockage is an altered exhaust note, confirmed by a glance at the exhaust outlet. Regular inspection of the intake strainer will not prevent blockage occurring in heavily contaminated water but will stop a gradual build-up stopping the water flow.

The connections in water cooled exhaust systems should be easy to get and inspect. Corroded clips should be replaced before they become difficult or impossible to remove, or worse, allow leakage.

With a wet exhaust system there is always a danger of water running back into the exhaust ports where it can cause damage. Preferably, the exhaust pipe should fall from the engine to outlet but if this is not possible for the complete system the pipe should be configured so that the engine is not the lowest point. The silencer, if any, should be placed in the final downward bend, so that there no danger of it filling with water. A drain cock should be fitted in the lowest part of the system.

Any back pressure in the exhaust system means that the engine has to work harder to pump the exhaust gases out of the system. Water cooling the exhaust can result in an increase in back pressure, so make sure that pipe diameters are adequate.

DRY EXHAUST

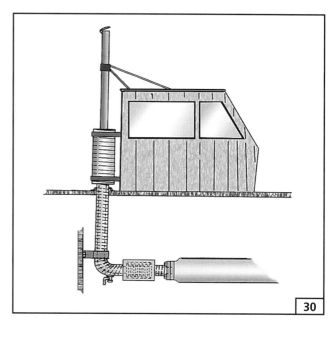

30

30. Open boats often have dry exhausts, the pipe running vertically upwards. This configuration tends to collect rain water and condensation so some sort of water trap and drain is recommended. A hinged flap over the end of the pipe will help to keep water out. Dry exhaust are open to the elements and often have flexible connections. Their useful life is generally shorter than water cooled exhausts, so inspection and replacement should be more frequent.

IMPORTANT NOTE: This type of exhaust system requires sufficient thermal lagging to reduce the risk of fire.

APPENDIX
TOOLS & EQUIPMENT

In this Appendix we mention some of the general-purpose workshop tools and equipment you will need, and we also provide details of specialist test and

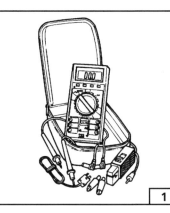

diagnostic equipment needed to accurately assess engine condition and faults.

General Workshop Tools

Although specialist tools and test equipment are needed to accurately diagnose engine condition and faults, their preparation and use often requires some basic hand tools. Specifically designed for undoing injection pipe unions, a split-ring spanner (also known as a flare nut wrench) is much better than an open-ended type of spanner.

Diesel Diagnostic Equipment

Depending on how deeply you are involved with diesel engine maintenance and tune-up, you may end up needing the majority of the following items of equipment.

ELECTRICAL SYSTEM TEST EQUIPMENT

Although a diesel engine does not have the troublesome ignition system necessary for petrol engines, the vast majority will have an electrical starter motor supplied from a battery and a battery charging system. Many other engines will also have an electrical circuit for cold starting devices.

BATTERY CONDITION TESTER

A battery tester may be needed to provide an accurate indication of battery condition. See **Chapter 3, Part C, Jobs 2 to 6** for more details on the rectification of battery and starter faults.

TEST LAMP

Connected between an earth/ground point or battery 'return' terminal and various parts of the electrical system, the test lamp will indicate the presence, or not, of battery voltage. The simplest, and cheapest item of electrical test equipment is a test lamp.

MULTI-METER

1. Because the multi-meter is much more than a PASS or FAIL device it has many more uses as a diagnostic tool. Always ensure it is preset and connected as described in the instructions. In particular, make sure that an appropriate scale is selected and that the cables are correctly attached.

CHECKING BATTERY CONDITION UNDER LOAD:

❑ **Step a.** Disconnect the STOP solenoid or activate the STOP mechanism, so that the engine will not run.

❑ **Step b.** Set the multi-meter (see *illustration 1*) to read volts, making sure the scale selected is suitable for the voltage to be measured.

❑ **Step c.** Connect the multi-meter across the battery terminals - positive lead (normally red) to the positive terminal and negative lead (normally black) to the negative terminal.

❑ **Step d.** Note the voltage reading, operate the starter and note the voltage with the engine being cranked.

❑ **Step e.** The expected reading will depend on engine size, battery size, starter motor type and engine temperature. As a guide, a 12V will show around 9V while the engine is being cranked. Expect a proportional drop for 24V systems. If the voltage reading is much below these figures, the starter motor current drain is too high and the starter motor should be repaired or renewed.

2. CHECKING STARTER TERMINAL VOLTAGE UNDER LOAD:

❑ **Step a.** First, check the battery-to-starter input voltage. Disconnect the STOP solenoid or activate the STOP mechanism, so that the engine will not run.

❏ **Step b.** Set the multi-meter to read volts, making sure the scale selected is suitable for the voltage to be measured.

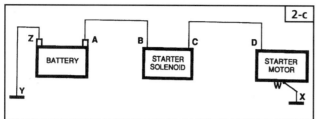

❏ **Step c.** Connect the multi-meter's positive lead to the positive battery terminal (**A**) and the multi-meter's negative lead to the starter input terminal (**D**).

❏ **Step d.** Operate the starter and note the voltage reading.

If the voltage drop is greater than 0.5V there is a high resistance in the starter supply circuit.

Carry out additional voltage checks between terminals **A** to **C** and **A** to **D** in illustration **2-c**, to find out where the problem lies. Ideally the voltage drop between the battery and the starter motor input terminal should be very low, in the order of half a volt. The example here shows how to check the system if the overall voltage drop is higher than ideal, and 1V is used as an example. The voltage is checked at different points to find out where the high drop occurs.

A test result such as A to D = 1.0V may break down to:
A to C = 0.9V
A to B = 0.2V - which indicates a problem between points B and C, the starter solenoid.

Confirm the location of the fault by connecting the voltmeter between B and C.

The connections should be checked and cleaned if necessary, or the solenoid should be replaced.

The second check is the battery to starter housing, point W (earth/ground return).

❏ **Step a.** Connect the positive multi-meter lead to the starter housing and the negative lead to the battery negative terminal.

❏ **Step b.** Operate the starter and note the voltage reading.

If the voltage drop is greater than 0.25V, there is a high resistance in the starter earth/ground return circuit. Carry out voltage checks between points W to X, X to Z and Y to Z (see illustration **2-c**) to isolate the problem

CHECKING CRANKING CURRENT:

Starter motors currents are relatively high, in the order of 400 - 500 amps, and only a few multi-meters are capable of measuring this current. A suitable multi-meter may use an inductive sensor on one of the battery cables to measure the starter motor current.

❏ **Step a.** Disconnect the stop solenoid or activate the stop mechanism so that the engine will not run.

❏ **Step b.** Connect the inductive clamp to either of the battery terminals.

❏ **Step c.** Crank the engine and note the current on the meter display.

❏ **Step d.** Compare the reading with the manufacturer's specification.

If the current exceeds the specification, the starter motor should be removed for repair or replacement.

CHECKING CHARGING VOLTAGE:

The multi-meter can also be used for checking the charging system voltage.

❏ **Step a.** Connect the voltmeter across the battery terminals.

❏ **Step b.** Start the engine and run it at 2-to-3 times idle speed, noting the voltage reading. A 12V system should give an indication of 13.8V to 14.8V. A 24V system should show similar margins over 24V.

❏ **Step c.** If the figures are low, check the alternator drive belt and repeat the tests. A high reading, or low reading after checking the drive belt tension, indicates voltage regulator problems. Have the regulator and alternator checked or replace them with new or serviceable used components. Prolonged running with an excessive charging system voltage will reduce battery life.

TACHOMETER

A tachometer is essential for checking and setting maximum engine speed. If an engine is difficult to start, the tachometer will also indicate whether the cranking speed is low. See **Chapter 3, Part C, Job 7** for details on checking cranking speed and **Chapter 4, Part C, Job 116** for details on maximum speed setting.

GLOW PLUG TESTER

While a multi-meter may be used to check glow plugs, and their control circuits, a dedicated glow plug tester will provide a more accurate diagnosis of the system.

3. The Dieseltune DX900 glow plug tester can be used to check the glow plugs either in or out of the engine. The instructions supplied with the tester provide specific details for testing
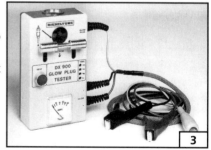
glow plugs and the test results should be compared with the manufacturer's specifications.

TESTING GLOW PLUG TIMER CIRCUIT:

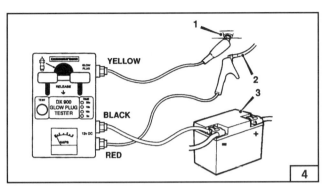

4. The Dieseltune DX900 glow plug tester can also be used to test the operation of the glow plug timer circuit in the following way:

☐ **Step a.** Make sure the engine is switched OFF.

☐ **Step b.** Disconnect the electrical supply lead from the controller to the glow plugs (at the glow plug end).

☐ **Step c.** Connect the red tester lead (**2**) to the disconnected glow plug lead, the black lead to the battery (**3**) negative post and the yellow lead to the disconnected glow plug (**1**).

☐ **Step d.** Press the red TEST button on the tester.

☐ **Step e.** Turn on the engine 'ignition' switch, watch the ammeter on the tester, and note the time taken for the controller to switch off. Note - the glow plug warning light will go out before the controller switches off.

☐ **Step f.** Compare the actual time with the manufacturer's specification. If the time is shorter than specified, engine cold starting may be difficult. Replace the controller if the engine is difficult to start.

Fuel Supply and System Testers

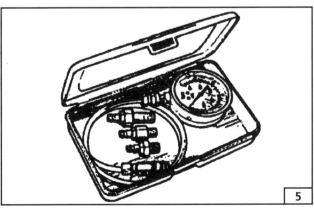

5. The fuel supply system can be checked with both vacuum and pressure gauges. The vacuum type fuel supply tester can

diagnose some injection fuel pump problems and blockages in fuel supply systems that do not have a lift pump.
The vacuum-type Dieseltune DX720 fuel supply tester...

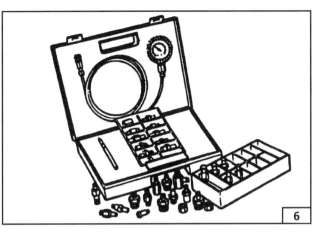

6. ...and the DX740 fuel system tester (combined vacuum and pressure type) are both supplied with a range of adapters to suit different engines.

CHECKING INJECTION FUEL PUMP:

☐ **Step a.** Read the instruction supplied with the fuel pump tester.

☐ **Step b.** Disconnect the fuel pipe at the injection pump.

☐ **Step c.** Fit a suitable adapter to the pump inlet.

☐ **Step d.** Make sure the pump is primed with fuel and connect the vacuum gauge to the adapter.

☐ **Step e.** Crank the engine and watch the vacuum gauge.

• If a vacuum is present, the pump is rotating.
• If no vacuum is seen, the pump or pump drive is faulty.
• Check the pump drive or remove and overhaul the pump.

BLOCKED FUEL SUPPLY

A fuel supply tester can be used to diagnose blocked fuel supply lines on engines that do not have fuel lift pumps.

☐ **Step a.** Disconnect the fuel supply pipe from the filter outlet and connect the banjo adapter supplied with the kit.

☐ **Step b.** Reconnect the fuel supply pipe and connect the vacuum gauge to the adapter.

☐ **Step c.** Start the engine and run at 2-to-3 times idle speed. Note the reading on the vacuum gauge.

If the vacuum is greater than 0.3 bar (9 in. Hg) suspect a choked filter or restricted fuel supply pipe between the filter and tank. The filter can be checked by fitting the tester to the filter inlet and repeating the test.

If the vacuum readings between filter inlet and filter outlet vary by more than 0.07 bar (2 in. Hg) the filter element is blocked and should be renewed.

FUEL SYSTEM TESTER

7. The Dieseltune DX740 is a fuel system pressure and vacuum tester that can be used to diagnose fuel lift pump problems and system restrictions.

❏ **Step a.** Disconnect the fuel inlet pipe at the injection pump and fit a 3-way adapter.

❏ **Step b.** Refit the fuel inlet pipe and connect the DX740 gauge.

❏ **Step c.** Start the engine and make sure there are no leaks.

❏ **Step d.** Increase engine speed up to maximum governed speed while watching the gauge.

• If pressure decreases as engine speed rises, suspect a faulty lift pump or a blockage in the supply line.
• Relocate the DX740 gauge to the lift pump outlet and repeat the test. If pressure is maintained, the filter or supply lines between the lift pump and injection pump are blocked.
• Clear or replace the fuel lines or renew the filter.
• If pressure still decrease as engine speed rises, relocate the gauge to the lift pump inlet and repeat the test.
• A vacuum reading that falls towards zero as engine speed increases indicates a faulty lift pump.
• An increasing vacuum indicates a blockage between the lift pump and fuel tank.

FUEL RETURN SYSTEM

The Dieseltune DX740 fuel system tester can also be used to check the condition of some fuel return systems where the manufacturer specifies fuel return pressures such as with the Lucas/CAV DPS and the Lucas/CAV DPC pumps.

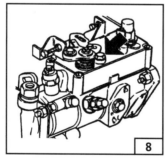

8. DPS PUMP: ❏ **Step a-1.** Remove the cam box vent screw (arrowed).

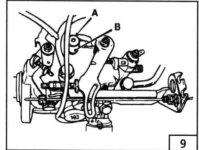

9. DPC PUMP: ❏ **Step a-2.** Remove the threaded timing pin plug (**A**) while restraining connector (**B**).

❏ **Step b.** Using a suitable adapter, fit the DX740 to the pump and start the engine.

❏ **Step c.** Watch the pressure gauge and increase engine speed. Note the maximum pressure reading:

• Cambox pressures between 0.6 and 1.15 bar (9 to 17 psi) are satisfactory.
• A pressure below 0.6 bar indicates a faulty cambox pressurising valve and the pump should be overhauled.
• If the pressure is above 1.15 bar, carry out the following test:

❏ **Step a.** Disconnect spill return pipe at pump outlet.

❏ **Step b.** Fit a length of pipe to the pump outlet and allow fuel to drain into a container.

❏ **Step c.** Start the engine and watch/note the maximum pressure reading:

• A cambox pressure above 1.15 bar indicates a faulty cambox pressurising valve.
• Pressures between 0.6 and 1.15 bar indicate a restricted or blocked return pipe.

AIR LEAKAGE TESTING:

Disconnect and seal the fuel supply line as near to the tank as possible. Alternatively, use fuel tank cocks to shut-off the fuel supply.

10. The Dieseltune Mityvac DX760 test kit can be used to test the fuel system for air leaks. Disconnect the fuel injection pump inlet pipe at the injection pump and use a clear plastic pipe to connect it to the Mityvac reservoir.

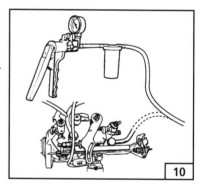

❏ **Step a.** Turn the Mityvac switch to VACUUM and operate until fuel is entering the reservoir. Continue to operate the tester until the gauge reads -0.7 bar.

❏ **Step b.** Stop pumping and watch the gauge.

• If there is a steady reading, there are no leaks.
• A falling reading indicates that air is entering the system.
• Repeat the test working back towards the fuel tank.
• Once the section of the system with the leak has been found, turn the switch on the Mityvac to PRESSURE and operate it until the system is pressurised to around 0.5 bar (7 psi). At this pressure, fuel should leak out, indicating the source of the leak.

STATIC ENGINE TIMING:

Ideally the timing should be checked with the engine running so that account is taken of any wear or backlash in operating mechanisms. However, the equipment needed for dynamic timing tests is expensive and not always immediately available. In many cases a static timing check will suffice and it will certainly identify any serious timing problems.

The techniques and equipment used for static timing depend on the injection pump type and the manufacturer's specifications. The simplest check is to align the various timing marks on drive wheels and pumps. Often the manufacturer specifies a particular size or type of pin that should be inserted to prevent rotation while the check and any adjustments are carried out.

SPILL TIMING - IN-LINE PUMPS ONLY:

11. Remove the injector pipe and delivery valve from the relevant cylinder outlet and substitute it with a short open pipe - a spill pipe. Turn the engine in the direction of rotation until fuel emerges from the spill pipe, then back slowly until fuel ceases to drip. This identifies the injection point quite accurately.

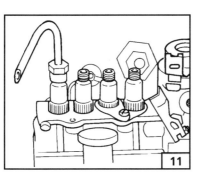

DIAL GAUGE TIMING:

12. Dial gauges are not cheap but they are very accurate and they do have other uses around an engine, such as checking valve opening points and valve lift. It is usually necessary to use an adapter. In this instance the dial gauge simply measures the point at which the plunger in the pump starts to move.

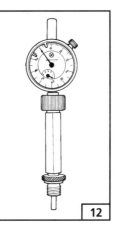

DYNAMIC ENGINE TIMING:

13. Checking the injection timing with the engine running is the only way to ensure accurate fuel injection. A piezo-electric point-of-injection pick up is clamped over the injector pipe of cylinder No 1 to detect injection.

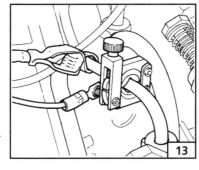

COLD IDLE ADVANCE:

Cold idle advance systems can be mechanically or electrically operated and the procedure for checking varies, depending on the type of system fitted.

MECHANICALLY OPERATED COLD IDLE ADVANCE:

☐ **Step a.** Prepare and connect a diesel timing pickup in accordance with manufacturer's instructions.

☐ **Step b.** Disconnect the linkage from the cold idle advance device.

☐ **Step c.** Start the engine and allow it to reach normal operating temperature.

☐ **Step d.** Check that the idle speed is correct and note the dynamic timing figure.

☐ **Step e.** Manually operate the cold idle advance lever and record the dynamic timing figure.

• An increase of 4 to 6 degrees is normal.
• If no change is recorded, the cold idle advance is faulty. The pump should be removed for overhaul.

ELECTRICALLY OPERATED COLD START ADVANCE:

☐ **Step a.** Prepare and connect the diesel timing system to be used, in accordance with manufacturer's instructions.

☐ **Step b.** Make sure that the engine coolant temperature is below 20 degrees Celcius, start the engine and allow it to idle.

☐ **Step c.** Record the dynamic timing figure.

☐ **Step d.** When the engine has reached normal operating temperature, check the idle speed and adjust if necessary.

☐ **Step e.** Record the dynamic timing figure.

• With the engine warm, the dynamic timing should be 4 to 6 degrees less than when cold.
• If there is no change as the engine warms up, the cold start advance electrical circuit should be checked.

Injector Testing

Although it is possible to carry out injector setting pressure and back-leakage tests with the injectors fitted, it is better if the injectors can be removed for a thorough test and physical examination.

IN-SITU TESTING:

14. The Dieseltune DX710 injector tester can be used to carry out limited tests on injectors while they are fitted to the engine. Make sure the tester is filled with calibration fluid.

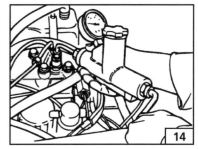

☐ **Step a.** Disconnect the fuel supply pipe to the injector and fit the injector tester.

☐ **Step b.** Operate the tester and compare the readings obtained with those specified by the manufacturer.

IMPORTANT NOTES:
• Only use suitable calibration fluid for testing. It is less hygroscopic (water absorbing) than diesel fuel and less harmful to the skin.
• Only test long enough to obtain the necessary readings. Lengthy tests can lead to combustion chamber flooding with a risk of knocking or hydraulic lock during restarting.
• Before running the engine, disconnect the injector and crank the engine on the starter motor to get rid of any excess liquid in the combustion chamber.

INJECTOR TESTING - REMOVED FROM ENGINE:

Injector valve opening pressure, spray pattern, dry seat (i.e. no leaks below the valve opening pressure) and back leakage (the rate at which injector pressure leaks back into the pipework) can all be checked with the injectors removed from the engine. Additionally, the physical condition of the injectors can be examined.

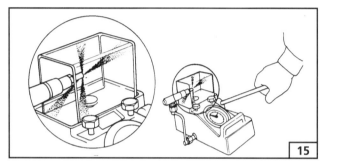

15. There are several injection testers available, of which the Dieseltune 111 tester and Hartridge types are good examples. Testers should be operated in accordance with the manufacturer's instructions, especially those relating to injector mounting.

SAFETY FIRST!

• *Diesel fuel is injected at very high pressures, so wear appropriate eye protection and do not let the fuel spurt or spray onto any part of the body.*
• *The pressures are high enough to force the fuel through the skin into the bloodstream with possibly fatal consequences.*
• *In particular, some modern common rail injection systems work at extremely high pressures - e.g. 1,350 bar in the Fiat and Mercedes Benz automotive systems.*
• *Leave very high-pressure systems to stand, without the engine running, for the period of time recommended by the manufacturer. This allows the extremely high pressure to be reduced, before attempting to depressurise the system in the normal way.*

Once injectors are removed, any deposits can be cleaned off with a **BRASS** brush. Holes can be cleaned with **BRASS** wire. Do not use steel brushes or wire.

ENGINE CONDITION:

A knowledge of engine condition can eliminate many possible causes of poor running. Cylinder compression tests will identify:
• poor inlet/exhaust valve sealing,
• worn bores/piston rings,
• a cracked cylinder head, or
• a leaking head gasket, but it will not identify which fault is causing the low compression. If a cylinder leakage tester is available it is usually possible to pinpoint the cause of poor compression.

CHECKING COMPRESSIONS:

16. Always use a compression tester suitable for use with diesel engines, such as the Dieseltune DX511 shown here. This type can be used to test the engine under operating conditions, i.e. at normal temperatures and running speeds,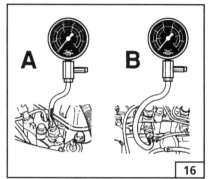
which is much better than testing at cranking speeds where the results can be affected by battery condition. With the adapters supplied, the tester can fitted either in the injector (**A**) or glow plug (**B**) locations.

SAFETY FIRST!

• *If the compression tester is fitted to a glow plug location, make sure the fuel supply to the injector is disconnected.*
• *Many compression testers are not designed to withstand combustion pressures and will be damaged if the cylinder fires while connected. Check manufacturer's instructions.*

Compare the test results with the manufacturer's specifications. Generally, any cylinder showing more than 15% lower than the highest reading is suspect. Since poor inlet valve sealing may be due to incorrect valve clearances, check them and re-test before going further.

CHECKING CYLINDER LEAKAGE:

A cylinder leakage test can pinpoint problems to a much greater extent than a compression test.

FACT FILE: PISTON POSITION
• Remember that the cylinder under test will have to have its piston at TDC after the compression stroke.
• If it is at any other position, either one or both valves may be open and this would obviously affect the compression/leakage readings.

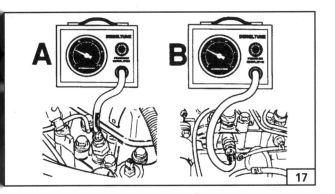

17. The Dieseltune DX520/DX540 is a typical cylinder leakage tester. Before use, the maker's instruction particularly those relating to safety, should be read, and the tester prepared for use. It can be connected to the injector (**A**) or glow plug (**B**) location.

❏ **Step a.** Remove the coolant filler cap.

❏ **Step b.** Connect the tester, either to an injector or a glow plug location.

❏ **Step c.** Operate the tester and note the readings. The following figures are a general guide to those expected from a warm engine:

	Car/LCV- size engines	Larger engines
Good	90%	80%
Marginal	80-90%	60-80%
Poor	Below 80%	Below 60%

• In general, larger engines are more tolerant of leakage.

• If the readings are marginal or poor, listen for air escaping:

AIR FROM:	INDICATES:
Air filter	Inlet valve leakage
Exhaust pipe	Exhaust valve leakage
Rocker cover	Worn piston rings/cylinder bores
Dipstick tube	Worn piston rings/cylinder bores
Radiator	Leaking cylinder head gasket

IMPORTANT NOTES:
• Although it may be possible to hear air leaking from the radiator cap, it is quite likely that the coolant level will rise before any air escapes.
• Air leaking past a valve may actually have come from an adjoining cylinder via a leaking head gasket.
• There will always be some leakage past piston rings.

Strip-down and Rebuild Tools

The following list is a small example of some of the specialist tools recommended by when work on an engine goes beyond routine maintenance.

18. Valve spring compressors of the lever type shown here are particularly useful for valve spring removal on diesel engines without having to remove the cylinder head.

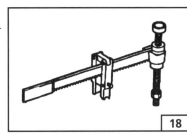

19. A piston height, valve depth and cylinder flange gauge is one of those tools that makes a job easier but it is intended for use with a dial gauge so has the potential for being an expensive combination.

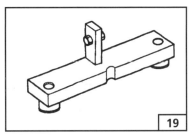

You should consult the manufacturer's manual for specific set-up and measuring details. With a little ingenuity and the use of feeler gauges and straight edges, this may not be an essential tool.

20. Pump timing is made easier with a universal timing gauge which usually has to be used with a drive adapter, gear adapter, pointer and, for Bosch pumps, distance piece.

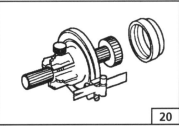

You should consult the manufacturer's manual for specific instructions.

21. A special adapter is available which can be bolted onto the crank-shaft pulley or flywheel of some engines to allow rotation via a 1/2 in. square drive. This is extremely useful for turning engines against

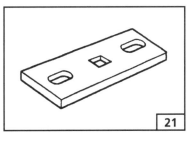

the high compression ratios found on diesel engines. It is a simple tool that could easily be made up in even a very basic workshop.